"Jack Pun is one of those rare English Medium Instruction personal expertise in both language and content. This dual expertise has enabled him to provide a tightly focused and authoritative contribution to the field."

Ernesto Macaro, *Emeritus Professor of Applied Linguistics, Department of Education, University of Oxford, United Kingdom*

"Jack Pun has carried out a detailed, systematic study of EMI practices in secondary schools in Hong Kong, focusing on both classroom interaction and teachers' and students' perceptions. The result is a rich and insightful analysis that will have wide-ranging implications for EMI policy and practice as well as medium-of-instruction research in diverse cultural, linguistic and educational contexts. It is a highly important contribution to the growing body of literature on EMI worldwide."

Li Wei, *Professor, Director & Dean, University of College London (UCL), Institute of Education, United Kingdom*

"Motivated by personal experience and based on rigorous empirical studies, this book provides an excellent account of the implementation of EMI in Hong Kong's secondary classrooms. This is a particularly valuable contribution to the EMI field because it offers an exemplary in-depth analysis of a specific context. There is a great deal to learn from this book, and I strongly recommend it to all those involved in the implementation of EMI."

Andy Kirkpatrick, *Professor Emeritus, Griffith University, Brisbane, Australia*

"A fascinating exploration of the issues teachers and students face when teaching and learning science through English Medium Instruction."

Jack C Richards, *Honorary Professor of TESOL, University of Sydney, Australia*

"Jack Pun deserves commendation for presenting to readers such a fine volume that has served to fill the gap in the existing literature. His focus on English as a medium of instruction in secondary school science classrooms, exploring issues relating to policies, practices and possible challenges, will compel readers to think deeply about this topic. The book will be a landmark work that researchers, practitioners and graduate students will find immensely useful."

Lawrence Jun Zhang, *Professor and Associate Dean (AD), International Partnerships Faculty of Education and Social Work, University of Auckland, New Zealand*

"This is a much-needed monograph on the implementation of English as medium of instruction in secondary schools. The monograph not only includes an in-depth analysis of classroom interactions, but it also focuses on the perceptions of students and science teachers. The findings have critical implications for teachers, teacher educators and policymakers to bridge the gap between policies and practices."

Xuesong (Andy) Gao, *Professor, School of Education, University of New South Wales, Sydney, Australia*

"*English Medium Instruction in Secondary Education: Policy, Challenges and Practices in Science Classrooms* by Jack Pun and Jack Richards is an essential read. It provides a deep and critical analysis of English as a medium of instruction, focusing on science classrooms. The book masterfully dissects policies, classroom interactions, and student experiences, and bridges the gap between theory and practice. A must-read for those shaping the future of EMI education."

Angel M. Y. Lin, *Chair Professor of Language, Literacy, and Social Semiotics in Education, Department of English Language Education, The Education University of Hong Kong, Hong Kong SAR, China*

"This book provides a rich and deep exploration of classroom discourse as well as teachers' and students' construction and perceptions of their EMI experiences in Hong Kong secondary school science classrooms. With its in-depth treatment of EMI policies, practices and challenges, the book offers insightful implications for stakeholders of EMI and is a must-read for policymakers, frontline teachers and researchers involved or interested in EMI in Hong Kong and beyond."

Guangwei Hu, *Associate Dean of Faculty of Humanities and Professor, Department of English and Communication, Hong Kong Polytechnic University, Hong Kong SAR, China*

"Not only will this volume entice all those involved in EMI science classrooms, but it will also appeal to EMI stakeholders in general, because the guidelines and reflections presented in order to deal with the language challenges found in classroom interaction are applicable to any EMI subject and context."

David Lasagabaster, *Professor of Applied Linguistics, University of the Basque Country UPV/EHU*

"The book provides a meticulous account of the nature and complexity of classroom interaction in EMI classes in secondary education as well as of teachers' and students' perceptions of their teaching and learning experiences. EMI practitioners, scholars, and policymakers interested in the interaction between policy and practice will find this book particularly useful. The theoretical perspectives on closing the gap between policy and practice are of exceptional value for the advancement of the field."

Slobodanka Dimova, *Professor in Multilingualism and Language Testing, University of Copenhagen, Denmark*

"This book presents a very comprehensive research study on English Medium Instruction (EMI) in Science classrooms. Contextualised in the Hong Kong EMI context, Jack Pun examines the interaction patterns in early-full and partial-late EMI Science lessons, as well as teachers' and students' perceptions. The quantitative and qualitative analyses of classroom interaction patterns not only provide interesting insights into the teaching and learning process in EMI lessons, but also illuminate classroom discourse research. With macro- and micro-level analyses and stakeholders' perspectives, this book depicts a clear picture of how EMI is implemented and perceived, thereby having implications for researchers, policymakers and teachers working in EMI and other bilingual education programmes."

Yuen-Yi Lo, *Associate Dean (Learning and Teaching) and Associate Professor, Faculty of Education, The University of Hong Kong*

"*English Medium Instruction in Secondary Education* makes a significant contribution to the field of EMI education. The book stems from Pun's personal journey in learning Science as a student growing up in Hong Kong. It offers policymakers and curriculum planners with insights from teachers and students on the opportunities and challenges on teaching and learning in the EMI Science classroom. Researchers can draw inspiration from the practicable approaches used to analyse classroom interactions with quantitative and quantitative methods. Teacher educators will also benefit from the thoughtful implications on professional learning for science teachers in EMI contexts. Overall, Pun has produced an excellent resource—one which is both valuable and timely for the community."

Victor Lim, *Associate Professor, English Language & Literature, National Institute of Education (NTU), Singapore*

"Jack Pun's personal narrative in the introductory sections offers compelling insight into the language barriers EMI learners encounter. Complemented by extensive theoretical discussions on EMI and research on classroom interaction patterns in EMI secondary science classrooms, this book unveils the complexities of implementing EMI in Hong Kong. It highlights the obstacles EMI learners face and outlines strategies for the professional development of EMI science and language teachers, making it essential reading for educators, researchers, and policymakers involved in the global EMI phenomenon."

Marie Yeo, *Senior Language Specialist, SEAMEO Regional Language Centre, Singapore*

English Medium Instruction in Secondary Education

Jack Pun presents best practices in pedagogy and teaching to facilitate effective content-subject learning at the secondary school level.

Increasingly, parents are sending their children to English Medium Instruction (EMI) secondary schools in their home countries, to prepare them for full immersion in EMI in English native-speaking countries. This book explores the teaching and learning processes in EMI senior secondary science classrooms based in thirty secondary schools in Hong Kong. Conducting analyses of classroom, teacher and student perception data, the author discusses the issues of teaching science through the medium of English in secondary schools, the implications and applications for professional development of science teachers and other content-subject teachers, and suggests strategies for teaching science in different EMI contexts.

This volume is highly relevant to scholars in the field of educational linguistics, particularly in English language teaching, content-based instruction, content and language integrated learning, and English as a medium of instruction. It is also useful to education policymakers, school teachers, research students, and English and education majors.

Jack Pun is an Assistant Professor in the Department of English at the City University of Hong Kong. He completed his DPhil at the University of Oxford, which explored the teaching and learning process in EMI science classrooms, with a special focus on classroom interactions, use of codeswitching, and teachers' and students' views of EMI. His research interests lie in EMI and health communication. His research has been published in journals such as *ELT Journal, Language Teaching, RELC Journal, Journal of English for Academic Purposes* and *International Journal of Bilingual Education and Bilingualism*. He is associate editor of *Journal of Research in Science & Technological Education*, and published two books on EMI: *Teaching and Learning in English Medium Instruction: An Introduction* (with Jack C. Richards) and *Research Methods in English Medium Instruction* (with Samantha Curle) by Routledge.

Routledge Studies in English-Medium Instruction
Series Editors: Diane Pecorari and Hans Malmström

Routledge Studies in English-Medium Instruction brings together research demonstrating the impact of work from applied and educational linguistics on the phenomenon of EMI. While maintaining a focus on EMI, rather than other forms of content and language integration more broadly, the series looks at EMI from a broad perspective. The series showcases research on a range of topics closely related to EMI such as assessment, delivery of instruction, translanguaging, vocabulary, language policy, support for instructors, the transition from CLIL to EMI and the development of L1 academic literacy. The series acts as a focal point for the growing body of research on EMI and will be of particular interest to students and scholars in applied linguistics, TESOL, TEFL, higher education pedagogy, and language education.

The Evolution of EMI Research in European Higher Education
Alessandra Molino, Slobodanka Dimova, Joyce Kling, and Sanne Larsen

English as a Medium of Instruction on the Arabian Peninsula
Edited by Mark Wyatt and Glenda El Gamal

Policies, Politics, and Ideologies of English Medium Instruction in Asian Universities
Unsettling Critical Edges
Edited by Pramod K. Sah and Fan Fang

English-Medium Instruction Pedagogies in Multilingual Universities in Asia
Edited by Fan Fang and Pramod K. Sah

English Medium Instruction in Secondary Education
Policy, Challenges and Practices in Science Classrooms
Jack Pun

For more information on the series, please visit: www.routledge.com/Routledge-Studies-in-English-Medium-Instruction/book-series/RREMI

English Medium Instruction in Secondary Education

Policy, Challenges and Practices
in Science Classrooms

Jack Pun

Routledge
Taylor & Francis Group
LONDON AND NEW YORK

First published 2024
by Routledge
4 Park Square, Milton Park, Abingdon, Oxon OX14 4RN

and by Routledge
605 Third Avenue, New York, NY 10158

Routledge is an imprint of the Taylor & Francis Group, an informa business

British Library Cataloguing-in-Publication Data
A catalogue record for this book is available from the British Library

ISBN: 9780367431457 (hbk)
ISBN: 9781032716770 (pbk)
ISBN: 9781003001454 (ebk)

DOI: 10.4324/9781003001454

Typeset in Times New Roman
by codeMantra

To my Mom (Siu Wai) and many who left us during COVID-19…

Contents

Figures, tables and excerpt

Figures

Tables

Excerpt

Acknowledgements

My heartfelt thanks go to all the people who have helped me with the book in many ways. First, I would like to thank my Oxford supervisors, Professor Ernesto Macaro and Dr Jane McNicholl, for many hours of stimulating discussion and debate. I would like to thank my examiners—Professor Andy Kirkpatrick, Dr Robert Woore, Dr Judith Hiller and Dr Ann Child—for their useful feedback at different stages of my DPhil.

Special thanks are also due to Dr Yuen-Yi Lo, Dr Mairin Hennebry, Dr Graham Lock, and Professor Diana Slade for their constant advice have been so helpful over the years. I would also like to give credit to the teachers and students who participated in my study for their generous help and support.

I acknowledge the kind and generous funding from the Swire Educational Trust over the past three years. I especially thank Mr R.P. Cutler and Ms Nora Rynne, for their warm meals and blessings. Without their generous financial and emotional support, I could not have studied at Oxford and made my dream come true. I am also grateful to Katie Peace, Khin Thazin and their colleagues at Routledge for their assistance and guidance in preparing this book.

Lastly I would like to thank my Mom (Siu Wai) and Dad (Wing-choi), who provided unconditional love and mental support during several critical stages in my DPhil journey, and who encouraged me to pursue my dream, as well as my church mates for being sympathetic or stern exactly whenever required. I could never have managed without you all!

Transcription conventions

[	overlapping speech
=	latching, or no gap between utterances
-	cut-off speech
?	rising intonation
,	continuing intonation
.	falling intonation
:	elongated sound
CAPS	loud speech
> <	fast speech
underline	marked stress
###	unintelligible speech
italics	analysts' descriptions of non-verbal communication
bold	focus of analysis

Introduction

I would like to begin this book with a personal story. I am passionate about languages and well aware of the challenges of becoming fluent in more than one language. Having grown up in a bilingual city, Hong Kong, where English takes precedence over my mother tongue, Cantonese, I understand the importance of speaking fluent English. To maintain my fluency in English, I studied very hard. I thought this would enable me to pursue a career in science, but this was not enough.

I grew up in a traditional Chinese family and Cantonese was our mother tongue. In my local primary school, I learned almost everything in Chinese. My interest in chemistry blossomed in secondary school. I was engaged in learning and always curious about the scientific principles that explained my surroundings. I developed my scientific skills by studying and exploring the properties and relationships of different materials at the atomic level. My laboratory work enabled me to develop strong skills in observation and lab preparation, and I learned how to investigate my hypotheses and draw conclusions based on scientific evidence. However, I encountered many difficulties with language. When I joined a local secondary school, I had to change my mode of learning from Cantonese-based to all-English. At that time, I thought obtaining good grades in English was all I needed to obtain a degree in chemistry at a university where English was the only language of instruction. However, this was not the case. Along with many of my science classmates who graduated from Chinese-medium secondary schools, I struggled to complete lab reports, assignments, and examinations. Not only was it difficult to learn the jargon in English, but it was also difficult to communicate in scientific English, which is completely different from conversational English. For example, the procedure section of a lab report is written in the past tense and the passive voice, but the discussion section is written in the present tense and the active voice. The Chinese language has no tenses or voices. The science and English teachers at my university could not understand my problems with the language of science. The lack of language support in the science department kept me from expressing my ideas freely. I initially struggled with learning the scientific language in my three-year chemistry programme, but I eventually overcame my language challenges.

DOI: 10.4324/9781003001454-1

This personal experience motivated me to study English after I received my bachelor's degree in chemistry in 2009. In 2010, I completed an English minor with a significant number of extra credits, particularly in scientific English. In 2011, my undergraduate chemistry research was published in one of the top chemistry journals (*Advanced Synthesis & Catalysis*). In the same year, I graduated with distinction with a cumulative GPA of 4/4 from my master's programme in English Language Studies. I also wrote a dissertation on the scientific language in science textbooks for secondary schools in Hong Kong. These accomplishments proved that I could express my ideas fluently in scientific English. The research and training I received in my English Language Studies programme provided me with a solid foundation for my four-year PhD in Education at the University of Oxford. My DPhil thesis is an exploratory study of the teaching and learning of secondary science in English in Hong Kong. These experiences motivated me to write this book and to share my journey and ways of coping with language challenges when studying science in English. Perhaps many science students in Hong Kong and around the world are facing similar language challenges—language challenges that are not explicitly addressed in the English curriculum. Students, particularly those with limited English proficiency or limited experience with English as the medium of instruction (EMI), often have no idea of what language requirements are placed upon them. Many students have been left behind, perhaps including the less-capable Chinese-medium instruction (CMI) students in Hong Kong who are being 'encouraged' to adopt EMI under the revised language policy described below.

In 2010, the Hong Kong government introduced the 'Fine-tuning Language Policy' to allow the conversion of CMI to EMI. Many students who used to study in Chinese were now being instructed in English. These former CMI students were now taught content subjects, such as physics, biology, and chemistry, in English. Although CMI schools switched to EMI, and the textbooks and examination materials that they use are written in English, most teachers and students in CMI schools still use a great deal of Cantonese for oral communication in the classroom (Poon, 2013). As a result, while many students may be learning science in English, they are interacting with their teachers and peers predominantly in Cantonese. My own experience suggests that in such an environment, students may not be able to develop adequate English skills or receive the support from teachers they need for full EMI instruction, such as in many EMI university settings where Cantonese is not used for interaction. During the early phases of policy implementation, the introduction of EMI did not necessarily benefit less-capable students, perhaps at the expense of the interest and learning needs of many former CMI students (Chan, 2013).

Common sense suggests that students' comprehension is better if they are educated in their first language (L1). For this reason, the education systems in many countries use L1 as the medium of instruction (MOI). However, in some countries, due to historical, political, and socioeconomic circumstances, a second language (L2), usually English, is used as the MOI. In places like Hong Kong, Singapore, India, Malaysia, and some African countries, English, a colonial language under past British sovereignty, has been retained to some extent as the MOI. As of 2016, 88.9% of Hong Kong residents speak Cantonese as their daily spoken language, whereas 4.3% speak English (Census and Statistics Department, 2016). Students sitting for the Hong Kong Diploma of Secondary Education are required to achieve at least Level 3 in English Language to meet the minimum requirements for UGC-funded undergraduate programmes (Hong Kong Examinations and Assessment Authority, 2021). A Level 3 in English Language implies adequate knowledge and understanding of the curriculum and the ability to apply concepts and skills appropriately in a variety of familiar contexts (Hong Kong Examinations and Assessment Authority, 2011, p. 7).

The choice of MOI in Hong Kong has been a controversial topic (Evans, 2000). The language policies for primary and tertiary levels are quite clear: most primary schools are CMI and all government-funded universities adopt EMI. However, the MOI in secondary schools has been a controversial topic in Hong Kong for at least the past two decades (Evans, 2013; Poon, 2013), and the debate over MOI in secondary schools is related to Hong Kong's history as a British colony from the mid-nineteenth century until its return to Chinese sovereignty in 1997.

Over the last half-century, there have been three stages of MOI development (Chan, 2014). The first stage was the British government's laissez-faire policy from 1946 to 1997. During this period, about 90% of secondary schools claimed to be "EMI". This meant that textbooks were widely written in English while the amount of spoken English used varied greatly depending on the type of school and the proficiency of students and faculty (Bolton, 2011, p. 8). The second period was the compulsory mother-tongue policy implemented after the handover of Hong Kong to China from 1998 to 2009, which compelled all public sector secondary schools to adopt CMI as a standard. The third period has been the fine-tuning of the MOI policy since 2010, which has allowed individual schools greater autonomy in choosing their MOI for different subjects at senior secondary levels.

Many studies have examined the impact of MOI on the quality of classroom interactions (Chan, 2013; Lo & Macaro, 2012; Lo, 2014). In one study, Lo and Macaro (2012) examined the quality of teacher–student interaction in junior secondary classrooms. They compared the quality of second language (L2) teaching in humanities and science classes before the implementation of the MOI policy in 2010. Students in humanities and science classes in Grades 9 and 10 in EMI and MOI-switching schools (e.g., schools switching MOI from CMI to EMI) were recruited. The results showed that science lessons were relatively teacher-centred, and that the teacher–student interactions consisted of simple Initiation–Response–Feedback (IRF) sequences. Lo and Macaro suggested that these interactional patterns might result in significantly less talking by students, with frequent short turns. Teachers in science lessons were found to pay less attention to form-focused instruction (e.g., teaching grammar rules and vocabulary) in L2 English. However, Lo and Macaro's study was conducted before the implementation of the 'fine-tuning' MOI policy. Their findings may not accurately describe the current MOI situation in Hong Kong, where Chinese-medium schools have been encouraged to adopt EMI for teaching science. Hence, we do not know the linguistic implications of the fine-tuning policy on schools that have elected to switch to EMI for science classes.

Based on the above, this book attempts to address the gap in our understanding of the current MOI situation in senior secondary schools in Hong Kong. The study explores how teachers and students interact in the EMI classroom and seeks to understand how teachers and students perceive their EMI experiences. Thus, the study examines two of the main dimensions of MOI: (1) classroom interactions and (2) teachers' and students' perceptions of the teaching and learning processes. To understand the impact in schools that have chosen to switch their MOI, the study compares these dimensions in two types of EMI in secondary schools: 'early–full EMI' and 'late–partial EMI'. These refer to traditional EMI schools and CMI schools that switch to EMI in secondary Grade 9, respectively. Drawing on the author's research for his DPhil thesis at the University of Oxford, this book extends Lo and Macaro's (2012) study by using a similar research design to examine the quality of classroom interactions in a new context—senior secondary science classrooms (physics, biology, and chemistry).

The study recruited 545 Grade 10 and 11 science students and 19 teachers in 8 EMI schools. Thirty-three science lessons were observed, producing approximately 1,360 hours of video recordings. In each school group, the teacher–student interactions in early–full EMI and late–partial EMI secondary schools and in Grades 10 and 11 were compared. The following data were used in the analyses: interaction time, proportion and length of teacher and student talk, pedagogical functions of teacher and student talk, proportion of L1 and L2 use, and teachers' questions and feedback. The science content in the observed lessons was analysed using Mortimer and Scott's (2003) framework (see Chapter 3).

In addition, a questionnaire and semi-structured interviews were used to examine teachers' and students' perceptions of the teaching and learning processes in EMI science classrooms. Teachers were asked about their views on EMI teaching, their language awareness in science teaching, their choices of instructional language in the classroom, their personal views on the students' language challenges, and the strategies they use to manage these challenges. Students were asked about their self-concepts in science, that is, their perceptions or beliefs about their ability to do well in science (Wilkins, 2004). The language challenges and coping strategies employed by each group of teachers and students were identified and compared to determine whether the strategies were similar in early–full and late–partial EMI schools.

Chapter 1 begins with a survey of global language immersion models and language policies. It then reviews the development of EMI policy in Hong Kong over the past few decades and presents the rationale for the research reported in this book. In Chapter 2, a number of relevant studies related to English as the MOI are reviewed. This review illustrates the variety of contexts for which language policies are designed and demonstrates that a 'one-size-fits-all' policy is unworkable. The remaining chapters report on research adapted from the author's DPhil thesis—Chapter 3 opens up the research site and describes the details of the large-scale study that informs the themes of this book. Chapters 4 through 9 present the results of the analyses of the classroom data and the teacher and student perception data. Chapter 10 discusses the implications of the findings and their applications to the professional development of EMI science teachers and language teachers.

References

Bolton, K. (2011). Language policy and planning in Hong Kong: Colonial and post-colonial perspectives. *Applied Linguistics Review, 2*, 51–74.

Census and Statistics Department. (2023, December 7). *Use of Language by Hong Kong Population*. Retrieved from https://www.bycensus2016.gov.hk/en/Snapshot-08.html

Chan, J. Y. H. (2013). A week in the life of a 'finely tuned' secondary school in Hong Kong. *Journal of Multilingual and Multicultural Development, 34*(5), 411–430. https://doi.org/10.1080/01434632.2013.770518

Chan, J. Y. H. (2014). Fine-tuning language policy in Hong Kong education: stakeholders' perceptions, practices and challenges. *Language and Education, 28*(5), 459–476. https://doi.org/10.1080/09500782.2014.904872

Evans, S. (2000). Classroom language use in Hong Kong's English-medium secondary schools. *Educational Research Journal, 15*(1), 19–43.

Evans, S. (2013). The long march to biliteracy and trilingualism: Language policy in Hong Kong education since the handover. *Annual Review of Applied Linguistics, 33*, 302–324. https://doi.org/10.1017/S0267190513000019

Hong Kong Examinations and Assessment Authority. (2011). *Grading procedures and standards-referenced reporting in the HKDSE examination*. Retrieved from https://www.hkeaa.edu.hk/DocLibrary/Media/Leaflets/HKDSE_SRR_A4_Booklet_Jun2011.pdf

Hong Kong Examinations and Assessment Authority. (2021). *Entrance requirements for undergraduate programmes*. Retrieved from https://www.hkeaa.edu.hk/en/recognition/hkdse_recognition/local/

Lo, Y. Y. (2014). L2 learning opportunities in different academic subjects in content-based instruction – evidence in favour of 'conventional wisdom.' *Language and Education, 28*(2), 141–160. https://doi.org/10.1080/09500782.2013.786086

Lo, Y. Y., & Macaro, E. (2012). The medium of instruction and classroom interaction: Evidence from Hong Kong secondary schools. *International Journal of Bilingual Education and Bilingualism, 15*(1), 29–52. https://doi.org/10.1080/13670050.2011.588307

Mortimer, E., & Scott, P. (2003). *Meaning making in secondary science classrooms*. Buckingham: Open University Press.

Poon, A. Y. K. (2013). Will the new fine-tuning medium-of-instruction policy alleviate the threats of dominance of English-medium instruction in Hong Kong? *Current Issues in Language Planning, 14*(1), 34–51. https://doi.org/10.1080/14664208.2013.791223

Wilkins, J. M. (2004). Mathematics and science self-concept: An international investigation. *The Journal of Experimental Education, 72*(4), 331–346.

1 English medium of instruction around the world

1.1 Medium of instruction (MOI) models around the world

In bilingual or multilingual countries, L2 immersion is often seen as promoting the appreciation of different cultural groups (e.g., in Canada) and increasing mobility or advanced knowledge transfer in a culturally diverse community (e.g., in the European Union). In English-speaking countries, immersion programmes may potentially allow new immigrants to integrate smoothly into the mainstream English education system (e.g., the United States, the United Kingdom, and Australia).

According to Swain and Johnson (1997, p. 6), there are eight key elements of a prototypical immersion programme:

1 L2 is the MOI.
2 The immersion curriculum is parallel to the traditional curriculum that is taught through L1.
3 There is overt support for L1.
4 The immersion programme aims to induce learners' additive bilingualism, i.e., learning an L2 without sacrificing the learning of the L1 in non-Anglophone contexts.
5 Learners' exposure to L2 is largely restricted to the classroom in non-Anglophone contexts.
6 Learners in immersion classrooms have similar or limited L2 language proficiency.
7 Immersion teachers are normally bilingual, sharing the same L1 and L2 as their students.
8 The classroom culture reflects only the students' L1 culture, rather than recreating the L2 culture.

In addition to these shared elements, global immersion programmes have features that distinguish them from each other (Swain & Lapkin, 2005). These features include:

• The level at which immersion is introduced (primary, secondary, or tertiary level within the education system).

DOI: 10.4324/9781003001454-2

- The type of immersion programme (early immersion: at the start of students' formal education; early–mid immersion: when students are in Grades 4 and 5; and late immersion: when students are in Grades 6 and 7).
- The extent of immersion (full immersion with only L2, partial immersion with less than half the content subjects taught in L2 or equal immersion between L1 and L2). The ratios of L1 to L2 at different stages within an immersion programme.
- The commitment of teachers and their schools, the attitudes towards the L2 culture.
- The status of L2 in society (Swain & Johnson, 1997).

Although different immersion programmes have their motives, socioeconomic contexts, and manner of implementation, they share the same goal of developing learners' L2 language proficiency by using the L2 to teach content subjects rather than by teaching the L2 as a language subject. The following sections briefly review different L2 immersion models.

Immersion programmes first appeared in Canada in the 1960s. The aim was not only to increase students' cultural awareness but also to develop their L2 language proficiency in subject areas. According to Swain and Johnson (1997), a group of Anglophone parents in St. Lambert, Quebec, were worried that their children were not achieving high proficiency in French. This could have had a negative impact on their children's future and economic success in Quebec, where French is the official language. The parents urged their school to introduce a completely different language programme, French immersion, to teach French by progressively exposing the students to French as the MOI. The belief was that this immersion approach would promote 'additive bilingualism'. That is, the children would learn French as an L2 while maintaining their L1, English. In the late 1960s, many Anglophone communities introduced such immersion programmes across Canada, recognising the high levels of communicative competence and advanced language proficiency students could gain through immersion and the value of speaking French.

The additive bilingualism provided by French immersion proved that immersion is one of the most effective approaches to L2 acquisition (Swain & Johnson, 1997). The approach influenced L2 teaching methods and was later widely adopted in many foreign language teaching contexts. One example is teaching English as a foreign language in the United States through content-based instruction (CBI) in the 1970s. Unlike Canadian immersion programmes, CBI aims to help new immigrants quickly acquire the necessary English skills to study content subjects like science before beginning their mainstream education with native speakers of English. CBI has also influenced the teaching of English as a foreign language (EFL) around the world (Davies, 2003). Nevertheless, some critics have challenged the effectiveness of these immersion programmes for foreign language learning. Problems that have been identified include teachers' inadequate expertise and competence in language teaching, their lack of experience in teaching content knowledge in the target language, and the difficulty in finding sufficiently challenging activities that match students' age and cognitive levels to ensure student motivation (Van Deusen-Scholl & Hornberger, 2008).

In Europe, the preferred immersion model has been content- and language-integrated learning (CLIL), which is 'a dual-focused educational approach in which an additional language is used for the learning and teaching of both content and language' (Coyle et al., 2010, p. 1). Since the 1990s, CLIL has gradually been adopted in different parts of Europe (European Commission, 2000). It aims to increase student mobility throughout Europe by increasing their English competence in content-subject areas. CLIL is similar to the immersion programmes in English-speaking countries. The goal is for students to acquire the L2 through a natural learning process (Jäppinen, 2005), which fosters bilingualism. Although CLIL and immersion are often used synonymously in foreign language research, Lasagabaster and Sierra (2009) have noted that there are more differences than similarities between CLIL and immersion, and suggested that a clear-cut distinction was needed, as the terminological confusion could have substantial consequences, with teachers not clearly understanding the language learning objectives and outcomes of these two approaches.

Lasagabaster and Sierra (2009) identified differences between CLIL and immersion in terms of classroom language, teacher training, sociolinguistic contexts, teaching principles, types of teaching materials, and achievable linguistic objectives.

First, the language in a CLIL classroom is not spoken in the community; thus, learners only have exposure inside the classroom. This is different from immersion programmes in which the languages may exist in some form in the student's local environment. Second, compared with immersion teachers, a relatively low percentage of CLIL teachers are native speakers of the language they are teaching. Third, the vast majority of CLIL programmes are introduced at the secondary or tertiary levels, making them late immersion programmes, whereas many immersion programmes start at an early stage, in pre-school or at the primary level, making them early immersion programmes. The difference in the years of exposure to the foreign language means that CLIL and immersion programmes are not comparable. Fourth, CLIL students have usually received traditional foreign language teaching in their primary education, in which the L2 (often English) is taught as a subject. In terms of language objectives, immersion programmes aim to help learners achieve native-like L2 proficiency, and teaching materials are thus written for native speakers. CLIL programmes do not necessarily have the same goal, and thus their teaching materials are abridged to accommodate learners with limited language proficiency. Dalton-Puffer (2007) noted that in some CLIL classrooms, teachers paid special attention to developing students' subject knowledge and L2 communicative competence, whereas teachers in immersion programmes taught students the subject knowledge in the target language, with less focus on language learning. Finally, students in immersion classrooms are often immigrants with different L1s, whereas, in CLIL classrooms, students are local students sharing the same L1.

To summarise, the models of immersion, CBI, EMI, and CLIL can be placed on a continuum according to the degree to which they focus on language or content. Table 1.1 summarises the different types of immersion language learning

Table 1.1 Focus and outcomes of different learning approaches

Language focus ←--------------------------→ Content focus

	CBI	CLIL	EMI	Immersion
Teachers	Language teachers	CLIL teachers (language or subject teachers)	Subject teachers	Immersion subject teachers
Teachers' language proficiency	Native speakers	Bilingual teachers (same L1 and L2 as students), qualified to teach subject matter only	Bilingual teachers (same L1 and L2 as students)	Native speakers
The kind of language teachers and students are working on	Work on language through subject content	Work on the language of subject content	Work on subject content in English	No attention paid to language as teaching is done in another language
Learners	Immigrants preparing to migrate to a mainstream English education system	Local students who received traditional English teaching as a subject at the primary level, with the same L1 and L2 as their EMI teachers	Local students who received traditional English teaching as a subject at the primary level, with the same L1 and L2 as their EMI teachers	Local students preparing to study or work in a region where another foreign language is the common or official language
Learner's language environment	Dominated by L2, exposure to L2 in both school and local contexts	Dominated by L1, exposure to classroom context	Dominated by L1, exposure to L2 restricted to classroom context	Dominated by L1, but L2 could also be found in the community and outside classroom contexts
Classroom culture	Culture of the L2	Similar to local L1 community	Similar to local L1 community	Culture of the L2
Aims	To teach language	To teach content and language	To teach content subject	To teach content subject
The kind of knowledge teachers refer to	Language knowledge	Content knowledge and knowledge about the language	Content knowledge	Content knowledge
Assumption teachers make about the learning process	Language is learned in context through topics	Content and language are complementary to each other	Content is learned through English	Content is learned without much attention to language
Examples	CBI in the United States	CLIL in Italian classrooms	EMI in Hong Kong	Canadian immersion
Goals	To gain subject knowledge by demonstrating language ability in a foreign language	Learning the foreign language and content knowledge is a dual process and connected, parallel objectives in CLIL classrooms	Learners are expected to demonstrate a threshold level of English proficiency before they are admitted to EMI programmes	To gain proficiency of the foreign language as a natural process through exposure in the environment, additive bilingualism

Source: Adopted from Dale and Tanner (2012); Wannagat (2007).

(Dale & Tanner, 2012; Wannagat, 2007). The columns on the left-hand side indicate that CBI is more language-focused than CLIL, EMI, and traditional immersion programmes, shown on the far right-hand side of the table, are primarily content focused, and CBI teachers see foreign language proficiency as the natural product of the learning process (Davies, 2003). They aim to offer students specific aims to achieve as well as a purpose in writing, contrary to a process approach, in which student writing is mostly tangential instead of serving a specific purpose (Hyland, 2003, 2004, 2007; Snow & Brinton, 2023). EMI is similar to immersion programmes, but it differs in allowing students who have passed a threshold level of proficiency to study in full EMI classrooms. According to Pecorari and Malmström (2018), EMI does not mean involving English as just the language of instruction. Instead, the definition of EMI means involving a setting in which the English language is intended to be utilised for some or all of a designed set of instructional goals (Pecorari & Malmtröm, 2018). In addition, L1 is not normally permitted in the classroom, because schools and policymakers see the use of L1 or code-mixing between L1 and English as a poor teaching practice that reduces the students' opportunity to practice English. However, in reality, many teachers adopt a code-mixing strategy or translanguaging to accommodate students with poor English abilities, as many of the students have not attained the level of English needed to communicate effectively and to logically express what they have learned in the class (Poon, 2013; Probyn, 2009). There is no language ability assumption for students in traditional immersion classrooms because these immersion programmes start in the early stages (e.g., primary level) of the education system. In many immersion programmes, the amount of L2 used is progressively increased, along with L1 instruction, across the years of the programme (Fang & Liu, 2020; Pun & Macaro, 2018; Wang & Curdt-Christiansen, 2019).

1.2   EMI as a global phenomenon

Many British ex-colonial regions such as Hong Kong, Malaysia, Singapore, and India have adopted EMI in their mainstream school systems. Graddol (2006), Macaro (2018) and Macaro et al. (2018) noted that global institutions in many non-English-speaking countries have also started to offer EMI in their university degree courses to attract more international students and teachers as the enrolment number of home students is falling through changing demographics and that there are national cuts in higher education investment. The wide adoption of EMI courses is demonstrated by a recent report on EMI provision in 55 countries. The report found that nearly 79% of public universities and over 90% of private universities allowed or sanctioned the use of English as the MOI (Dearden, 2015).

In Europe, the Bologna declaration signed in 1999 aims to promote student and faculty exchange across the EU and to develop a common higher education framework, with comparable degrees and easy academic credit transfer (European Ministers of Education, 1999). A direct consequence of this declaration is the massive growth in the number of EMI courses offered at European universities (Coleman, 2006; Doiz et al., 2011). Studies have also reported a change in the national language policies in some European countries, with different interpreta-tions of EMI policy by different stakeholders (e.g. school management, lectur-ers, students). The challenges of implementing EMI courses in university settings have been studied in different European countries, including Austria (Tatzl, 2011), Denmark (Thøgersen & Airey, 2011), Italy (Francomacaro, 2011), Norway (Ljos-land, 2011), Sweden (Airey & Linder, 2008; Björkman, 2008), Spain (Doiz et al., 2014) and Switzerland (Studer, 2014).

A Bologna-like plan was announced at the 2012 Asia-Pacific Economic Co-operation (APEC) summit, which sought to explore ways of facilitating faculty and student mobility across Asia. This will inevitably speed up the adoption of the already widespread practice of offering EMI courses at Asian universi-ties. For instance, Malaysia adopted the Malaysia Education Blueprint in 2013, which covers post-secondary education and includes bilingual proficiency in Malay and English as an educational requirement (Ministry of Education Ma-laysia, 2013, p. 31). As of 2018, several HEIs in Vietnam have also developed programmes delivered entirely in EMI, both to help Vietnamese students gain English proficiency and accommodate foreign students (Asia-Pacific Economic Cooperation, 2019, p. 36). These EMI programmes aim to improve teachers' and students' English proficiency, increase university rankings, and attract more for-eign exchange students (Kirkpatrick, 2014a). Studies have examined the issues related to the expansion of EMI programmes at the university level in Africa (Alidou, 2004), the Asia-Pacific region (Kirkpatrick, 2011, 2014a; Lei & Hu, 2022), China (Hu et al., 2014; Rose et al., 2020), Hong Kong (Lin & Morrison, 2010), India (Annamalai, 2004), Korea (Byun et al., 2010; Kim, 2011), Malaysia (Othman & Mohd Saat, 2009), Singapore (Pakir, 2004), Thailand (Lavankura, 2013), Taiwan (Hou et al., 2013), Turkey (Demirbulak, 2011) and Vietnam (Dang et al., 2013).

A review by Hamid et al. (2013) explored the policy and practices of MOI in ten Asian regions (Bangladesh, Hong Kong, India, Indonesia, Japan, Malaysia, Maldives, Nepal, Timor-Lester, and Vietnam) and concluded that the dominance of EMI is a response to globalisation pressures and the need to increase students' English proficiency, as English is an international language. However, differences in the MOI practices and policies and the extent of EMI use in different regions reflect each region's distinct socioeconomic and educational motivations and their desire to promote their own identities, cultures, and first languages.

Although EMI is recognised as an important strategy for ensuring the compet-itive success of education systems in many non-English-speaking countries, con-cerns about the quality of EMI implementation have arisen in many countries. Problems include the lack of EMI training for teachers and poor English profi-ciency among students (Tamtam et al., 2012). In addition, Kirkpatrick (2014b) has pointed out that universities in Asia have not developed a language policy that enhances the quality of EMI. He argued that Asian universities do not consider the multilingual setting in which they are located and do not recognise the varie-ties of English that are spoken in the Asian region. For example, in Hong Kong, all eight government-funded universities are of English medium, in principle.

However, only two universities have some form of language policy that promotes bilingualism or trilingualism. The Education University of Hong Kong (EdUHK) has a clear language policy designed to promote functional trilingualism in Cantonese, Mandarin, and English (Kirkpatrick, 2014b). Full-time undergraduates at the university are subject to language exit requirements, which demand that they attain an IELTS score of 6.0 and a PSC score of 3B before graduation (The Education University of Hong Kong, 2021, p. 1). The Chinese University of Hong Kong (CUHK) has a bilingual policy (Cantonese and English), although this was threatened by a decision made by the previous president of CUHK to increase the number of EMI courses as part of the university's internationalisation process, designed to attract overseas staff and students, and improve its ranking. In response, in 2005, the students and alumni of CUHK staged a protest, calling for the president's resignation for what they perceived as mishandling the language issue. Subsequently, a student filed a legal case against the university's directive to deliver more EMI courses, charging that it demeaned the importance of Chinese and disregarded the original mission of CUHK. However, this legal challenge failed, and the Supreme Court of Hong Kong ruled that CUHK had no legal obligation to keep Chinese as the MOI because its founding ordinance did not mandate the university to use Chinese as the principal language of instruction (Kirkpatrick, 2014b). CUHK was founded in 1963 through the amalgamation of a number of previously Chinese-medium private colleges to provide Chinese-medium tertiary education. The decision to use spoken Cantonese and written Modern Standard Chinese as its MOI placed it in direct competition with the only other university in Hong Kong at that time, the University of Hong Kong, which offered English medium education (Lin & Man, 2011). The university's most recent report on bilingualism states that universal academic subjects that have little cultural specificity, such as science and engineering, should be taught in English. It also suggests that the choice of MOI should be flexible and should consider the nature of the academic subject; discipline requirements; students' language habits, competence, and cultural backgrounds; and teachers' concerns and practical needs. These decisions about the ratio of Cantonese, Putonghua, and English used for teaching at the university can now be made at the department level (Chinese University of Hong Kong, 2007). CUHK and EdUHK in Hong Kong present two good examples of localised language policies inside tertiary institutions that reflect the cultural setting in which they are located. This is related to Kirkpatrick's (2014b) suggestion that universities in Asia need to recognise the full potential of different 'Englishes' as a lingua franca when adopting EMI.

1.3   Language educational policies in Hong Kong

In Hong Kong, a former British colonial city, English is still an official language in the government, education, and business sectors. High proficiency in English confers superior socioeconomic status to the user and is an essential requirement for academic and career success (Evans & Green, 2007; Tsui, 2007). Parents in

Hong Kong understand that good English is the key to their children's success in a globalised world where English is the lingua franca (Poon, 2010). Although most primary schools are CMI and all government-funded universities adopt EMI, MOI at the secondary level has been a controversial topic, leading to a major public debate that has moved through four main stages over the past four decades (Poon, 2013).

Stage 1: In the 1960s, there were two types of secondary schools in Hong Kong: Anglo-Chinese schools using EMI and traditional Chinese schools using CMI (Education Commission, 1981).

Stage 2: In the 1980s, priority was given to teaching Hong Kong students both Chinese and English. Combined with the massive expansion in secondary education, this led to a new vision for Hong Kong. The aim became to not only produce a group of elite English users but also foster education among those for whom Chinese was the main medium of communication. Schools made the shift towards a complete L1 education in the early compulsory years (e.g., CMI). The first nine years of schooling (primary Grades 1–6 and junior secondary Grades 7–9) were taught in Chinese, with the exception of a few EMI schools that had proven to be successful in teaching content subjects in English. The distinction between Anglo-Chinese and Chinese secondary schools was abandoned in the late 1990s; each type of school could use only one language as its MOI.

Stage 3: In the 1980s and 1990s, many schools opting for EMI adopted a mixed-code teaching approach in practice to delivering the curriculum (Education Commission, 1990). Although teaching and exam materials were written in English, teachers and students interacted mainly in Cantonese, with supplementary English technical vocabulary. Teachers in these schools used Cantonese as the primary instructional language for teaching content subjects, supplemented by English technical vocabulary. English was used as the language for reading and writing in the classrooms. Some teachers spoke only Cantonese when teaching English scientific writing (Lin, 1997). Students used English textbooks and memorised English terminology but communicated with teachers only in Cantonese during discussions, matters of classroom discipline and relationship building (Lin, 2006; Yip et al., 2007). Such mixed-mode instruction was prevalent in Hong Kong's secondary schools in the 1980s and 1990s (Tsui, 2004).

The use of mixed-code instruction in EMI schools, in which teachers constantly used Cantonese rather than English to accommodate students' poor English proficiency, was heavily criticised by both the public and education sectors (Poon, 2013). For students, one disadvantage of this form of mixed-mode instruction was that teachers only used English to ask recall questions, which encouraged students to memorise what they needed for examination, leading to the rote memorisation of facts in English and time wasted on the translation of English texts. The rote memorisation of technical terms in English meant that this approach presented obstacles to the development of higher-order cognitive thinking (Education Commission, 1988, 1990, 1992, 1996).

In 1998, to avoid the claimed drawbacks of mixed-mode teaching and ensure there was only one language of instruction in a school, the government adopted what it considered to be a clear-cut language policy: all schools should adopt CMI, apart from a few exceptional schools with proven success in EMI teaching. All primary Grade 6 students were tested on their English proficiency before they were admitted to secondary schools. Only students who passed the threshold level of English competence could continue to use English as the MOI at the secondary level. In 1998, the number of EMI schools dropped significantly from approximately 90% to 25% (Glenwright, 2003). All secondary schools in Hong Kong were separated into either EMI or CMI schools. This policy allowed 114 secondary schools to adopt EMI for teaching content subjects (e.g., science, history, geography), and the remaining 307 schools used CMI. Upon finishing primary Grade 6 education, Hong Kong primary school students undertook the Academic Aptitude Index test and were classified into different secondary school MOI groups based on their performance.

During the eight years of its implementation, the 'clear-cut' mother-tongue policy was severely criticised by stakeholders (e.g., employers, parents, teachers, and students) for its contribution to the perceived decline in students' overall English proficiency. Criticism was vocal, even though government research suggested that CMI might provide learning benefits to both teachers and students by enabling diversified teaching modes, closer relationships between teachers and students, and effective classroom discussions. The government claimed that "CMI is bearing fruit" (Education Bureau, 2005, p. 8), and suggested that students could develop higher-order thinking more effectively in their L1, leading to better academic outcomes and student achievements at the junior secondary level (Education Bureau, 2005). However, criticisms linking CMI to unsatisfactory levels of English proficiency among students were a major factor behind a parent-led push urging a reversal of this compulsory policy in 2010 (Poon, 2013).

Stage 4: In 2010, the government 'fine-tuned' its previous language policy. The new policy aimed to improve students' English proficiency by providing an environment for English learning in non-language subjects through immersion. Under this initiative, secondary schools that were using Cantonese as their MOI but which met certain criteria were allowed to switch to EMI.

A recent review of the development of MOI policy found that the 'fine-tuning' policy and the use of Mandarin as the MOI in Chinese language teaching have added more points of concern to the current MOI situation by reducing the role of L1 instruction (Evans, 2013). The increasing adoption of Putonghua rather than Cantonese as the MOI for Chinese subjects has been attributed to the closer correspondence between the national language and standard written Chinese (Li, 2006; Snow, 2004). It is therefore considered a more effective means of enhancing Chinese literacy (Standing Committee on Language Education and Research, 2003). This revised policy gave individual schools, especially CMI schools, greater autonomy than the previous compulsory mother-tongue language policy, which mandated CMI at the primary level and limited the number of EMI secondary schools. Through this new policy, schools could decide on their preferred MOI (e.g., English, mixed-mode, or Chinese-only) for content subjects based on the student's learning ability, the teachers' capability, the requirements of individual subjects, and school resources. The government stressed that this fine-tuning policy maximised the educational value of both MOIs (Legislative Council, 2009). Students with poor English proficiency could learn effectively through L1 instruction in their early years, and those who have reached a threshold level of English could progress from partial to total immersion in English in subject areas. The implementation of this policy was also believed to reduce the labelling that arose from the segregation of CMI and EMI schools. The policy intended to ensure that students with poor English proficiency would not be discriminated against because they were attending a CMI school.

Under the fine-tuning language policy of 2010, CMI schools in Hong Kong started to implement EMI in senior secondary Grades 10–13 (Legislative Council, 2009; Poon, 2010). Students who were exposed to Chinese as an MOI at junior levels were expected to improve their English skills by adopting English for content subjects such as physics, biology, and chemistry. Through this policy, many CMI schools have now opted for more EMI classes for two reasons: (1) public pressure after evidence of a decline in the English proficiency of their previous graduates and (2) an increase in the strongly expressed parental desire for children to become fluent in English.

However, most students that adopt English for content subjects have limited English ability and no experience learning through EMI (Poon, 2013). The expectation that students' English ability will improve through EMI is problematic (Dearden, 2015). In content subjects like science, teachers and students used to a Cantonese environment have now switched to English medium, starting at senior levels (Grade 10) and extending to all levels. However, studies have pointed out problems with this 'late–partial' English immersion (Lin, 2006; Poon, 2004, 2009), including problems with students' inadequate English proficiency, especially their lack of technical vocabulary, which leads to the student's inability to understand the subject content knowledge or to express their ideas effectively in English (Yip et al., 2007). For content-subject teachers, the lack of training in delivering content in English has led to a decline in the quality of teaching in these classrooms (Chan, 2014; Kwan, 1989; Poon, 2013; Pun & Thomas, 2020). Thus, many teachers compensate for their students' limited English proficiency by using Cantonese as the main instructional language (Lin, 2006).

The educational benefits that these students receive from EMI are questionable. Chan (2014) has argued that the increase in EMI courses has benefited students who adapt well to rote learning and didactic teaching methods, at the cost of the interests and learning needs of less capable students. By interviewing principals, teachers, and students in late–partial EMI schools, he found that the increase in EMI teaching had resulted in many students encountering challenges related to language (e.g., poor English communication skills) and led to pedagogical and practical issues in content-subject classrooms (e.g., limited interaction with teachers).

Similar results have been found in schools that implement EMI at the junior secondary level. Poon et al. (2013) asked 461 Grade 7 secondary students about their EMI experiences. Although at least half of the students preferred to be placed in classes that were nominally EMI, the majority of these students did not seem to favour the use of English for teaching and learning purposes. They encountered numerous language difficulties when switching from Chinese to English, such as hindered understanding of content-based subject knowledge due to large quantities of vocabulary and complicated sentences. The difficulties were even greater for students in the late–partial EMI schools, where most of the students did not possess sufficient English skills for EMI classes.

As indicated by Chan (2014) and Poon et al. (2013), the educational values offered by the fine-tuning MOI decision appear limited. Due to the many language challenges that students encounter in EMI classrooms, such as problems understanding abstract concepts and integrating both content and language (Schleppegrell, 2004; Yip et al., 2003), it is questionable whether these CMI students, particularly those with poor English proficiency, can learn effectively in an English environment. However, the two studies that have examined the impacts of the fine-tuning language policy (Chan, 2014; Poon et al., 2013) provide very little information about teaching and learning processes that are crucial for learning, such as the nature of classroom interaction. The next section summarises the major findings from studies that have examined classroom interactions in EMI classrooms in Hong Kong.

1.4   What research tells us about language interaction patterns in Hong Kong

Studies of second language acquisition (SLA) have suggested that meaningful negotiation plays an important role in the L2 acquisition process. Such negotiations can be conducted successfully using the target language with guided feedback. According to the interaction hypothesis (Long, 1983), when learners are exposed to more comprehensible input, they can produce more L2 output through meaningful interactions with teachers, who can receive the students' modified input and provide corrective feedback on their L2 use, which ultimately contributes to students' learning.

Teacher–student interactions provide a social context in which learners can focus on and reflect on language forms and meanings while using them. It helps learners understand how the grammar of the language works in practice and how grammatical choices shape meaning. It is important that teachers should be strategic in their language use. Instead of merely using L1 when delivering teaching content verbally, teachers should be aware of their duty to encourage their students to make the most of their linguistic repertoire and switch tactfully between L1, everyday L2 as well as academic L2 (Lo, 2015).

The interaction hypothesis has been supported by research on EMI learning in content subjects in Hong Kong. These studies have shown that to develop both the required level of English proficiency and the linguistic capacity for L2 learning in content subjects, students need to fill in their current language gaps (Yip et al., 2003). Yip (2004) argued that efforts should be made to close the gap between how students conceptualised scientific knowledge in L1 and L2. This can be achieved by using prompting questions to evaluate students' scientific understanding and identify their misconceptions (Yip, 2004). The gap between what students can understand in English and what they can articulate in English can be closed by providing sufficient L2 learning opportunities through classroom interactions to test their L2 hypotheses (Lo, 2014) and by giving corrective feedback on their L2 utterances.

It is questionable whether such language and knowledge gaps can be closed in late–partial EMI secondary schools if teachers and students only interact in English. These students are instructed in CMI at the junior levels and then have to switch to EMI at the senior levels. As a result, they have limited English ability and perhaps no experience in EMI learning. This may result in a mixed mode of instruction in many late–partial EMI secondary schools.

Although research in SLA has suggested that language acquisition is possible in L2-only settings, a growing number of studies have validated teachers' use of L1 in content teaching, especially for teaching L2 vocabulary (Zhao & Macaro, 2014). There is also overwhelming evidence that most EMI teachers consider the L1 an indispensable tool in L2 pedagogy (Lin, 2006; Probyn, 2001, 2009; Pun & Macaro, 2018) since they think what is being communicated in L2 is real and important (or 'serious') information (Pun & Macaro, 2018). For example, Lin (2006) analysed teaching practices in Hong Kong science classrooms and proposed that drawing on the students' L1 and cultural resources could help junior CMI science students with poor English proficiency to use more technical terms in English.

EMI classes are predicated on a belief that massive exposure to L2 leads to L2 learning. However, many students with poor English ability cannot benefit from English-only classrooms. It is therefore perhaps appropriate to ask how L1 can play an important role in EMI teaching, given that a significant number of studies have reported the benefits of L1 in EMI teaching (Lin, 2006; Lo & Macaro, 2012; Macaro, 2020; Probyn, 2006). L1 can perhaps be beneficial in EMI contexts where content learning is the primary objective rather than language acquisition, yet there is very little research on L1 use in EMI settings (Lo & Macaro, 2015), and practical pedagogical suggestions for supporting students' language challenges appear inadequate (Chan, 2014; Poon et al., 2013). Some scholars researching Global English have suggested that EMI does not and should not follow 'native speaker' norms such as expectations of the use of monolingual 'English-only' (Sahan et al., 2022). The government and EMI schools should also not be promoting a monolingual form of education (Rose & Galloway, 2019). Instead, the use of multilingual and translanguaging practices should be recognised (Curle et al., 2021; Sahan et al., 2022). This is because former studies on translanguaging practices in education have proven that translanguaging is very common during students' interactions (Kuteeva, 2020), is utilised by teachers for a wide range of pedagogical goals (Mazak & Herbas-Donoso, 2015) and provides an alternative to English-only approaches to EMI teaching (Sahan & Rose, 2021). These pedagogical suggestions should be taken into account to improve the efficiency of EMI education.

This book is a response to these current gaps in the EMI literature. It sets out to explore how teachers and students interact in science classrooms, and how they feel about their EMI teaching and learning experiences. Thus, it compares teacher–student interactions in science classrooms in both 'early–full' and 'late–partial' EMI schools and investigates teachers' and students' perceptions of the EMI experience in these science classrooms. The research questions are as follows.

1 What is the nature of classroom interactions in EMI science classrooms?
2 What are the differences and similarities in classroom interactions in early–full and late–partial EMI schools?
3 What are the differences and similarities in classroom interactions between Grades 10 and 11 in early–full and late–partial EMI schools?
4 What are teachers' perceptions of teaching science through EMI?

1.5 Summary

The findings in this book can provide teachers with authentic practices to better teach science and recognise the challenges and problems in language use. It is only by using a language in the science classroom and unravelling the complexities of the language that students will be able to improve their ability to read and write scientific texts in that language. Linguistic analyses of classroom interactions are therefore important because they can provide teachers with information about which aspects of scientific language students need to improve. The goal of this book is to show how scientific knowledge is interpreted through language and to shed light on the need for the use of scientific English in the science classroom and in practice. The next chapter will situate the research question, critically assess previous research on the topic, identify existing knowledge gaps, formulate the research questions and argue for the need for this research project.

References

Airey, J., & Linder, C. (2008). Bilingual scientific literacy? The use of English in Swedish university science courses. *Nordic Journal of English Studies, 7*(3), 145–161.

Alidou, H. (2004). Medium of instruction in post-colonial Africa. In J.W. Tollefson & A. B. M. Tsui (Eds.), *Medium of instruction policies: Which agenda? Whose agenda?* (pp. 195–216). London: Routledge.

Annamalai, E. (2004). Medium of power: The question of English in education in India. In J.W. Tollefson & A. B. M. Tsui (Eds.), *Medium of instruction policies: Which agenda? Whose agenda?* (pp. 177–194). London: Routledge.

Asia-Pacific Economic Cooperation. (2019). *APEC 2018 Report on education and economic development*. Retrieved from https://www.apec.org/Publications/2019/11/APEC-2018-Report-on-Education-and-Economic-Development

Björkman, B. (2008). 'So where we are?' Spoken lingua franca English at a technical university in Sweden. *English Today, 24*(2), 35–41. https://doi.org/10.1017/S0266078408000187

Byun, K., Chu, H., Kim, M., Park, I., Kim, S., & Jung, J. (2010). English-medium teaching in Korean higher education: Policy debates and reality. *Higher Education, 62*(4), 431–449. https://doi.org/10.1007/s10734-010-9397-4

Chan, J. Y. H. (2014). Fine-tuning language policy in Hong Kong education: Stakeholders' perceptions, practices and challenges. *Language and Education, 28*(5), 459–476. https://doi.org/10.1080/09500782.2014.904872

Chinese University of Hong Kong. (2007) *Final report on bilingualism: Chinese University of Hong Kong.*

Coleman, J. A. (2006). English-medium teaching in European Higher Education. *Language Teaching, 39*(1), 1–14. https://doi.org/10.1017/S026144480600320X

Coyle, D., Hood, P., & Marsh, D. (2010). *CLIL: Content and language integrated learning.* Cambridge: Cambridge University Press.

Curle, S., Jablonkai, R.R., Mittelmeier, J., Sahan, K., & Veitch, A. (2021). Part 1: Literature review. In Galloway N. (Ed.), *English in higher education – English medium.* British Council. https://www.researchgate.net/publication/344352859_English_in_higher_education_-_English_medium_Part_1_Literature_review

Dale, L., & Tanner, R. (2012). *CLIL Activities: A resource for subject and language teachers.* Cambridge: Cambridge University Press.

Dalton-Puffer, C. (2007). *Discourse in content-and-language-integrated learning (CLIL) classrooms.* Amsterdam: John Benjamins.

Dang, T. K. A., Nguyen, H. T. M., & Le, T. T. T. (2013). The impacts of globalisation on EFL teacher education through English as a medium of instruction: an example from Vietnam. *Current Issues in Language Planning, 14*(1), 52–72. https://doi.org/10.1080/14664208.2013.780321

Davies, S. (2003). Content-based instruction in EFL contexts. *The Internet TESL Journal, IX*(2), 9–17.

Dearden, J. (2015). *English as a medium of instruction-a growing global phenomenon.* Retrieved from https://www.britishcouncil.org/sites/default/files/e484_emi_-_cover_option_3_final_web.pdf

Demirbulak, D. (2011). A look at the Turkish and English Language in Turkey from the perspectives of tertiary undergraduate students. *Procedia - Social and Behavioral Sciences, 15*, 4083–4088. https://doi.org/10.1016/j.sbspro.2011.04.419

Doiz, A., Lasagabaster, D., & Sierra, J. M. (2011). Internationalisation, multilingualism and English-medium instruction. *World Englishes, 30*(3), 345–359. https://doi.org/10.1111/j.1467-971X.2011.01718.x

Doiz, A., Lasagabaster, D., & Sierra, J. M. (2014). What does 'international university' mean at a European bilingual university? The role of languages and culture. *Language Awareness, 23*(1–2), 172–186. https://doi.org/10.1080/09658416.2013.863895

Education Bureau (2005). *Report on review of medium of instruction for secondary schools and secondary schools allocation.* Hong Kong.

Education Commission. (1981). *The Hong Kong education system.* Hong Kong.

Education Commission. (1988). *Education commission report no. 3* (Vol. 3). Hong Kong.

Education Commission. (1990). *Education commission report no. 4* (Vol. 4). Hong Kong.

Education Commission (1992). *Education commission report no. 5* (Vol. 5). Hong Kong.

Education Commission. (1996). *Education commission report no. 6* (Vol. 6). Hong Kong.

European Commission (2000). *European report on quality of school education. Sixteen quality indicators. Report based on the Work of the Working Committee on Quality Indicators.*

European Ministers of Education. (1999). *The Bologna declaration of 19th June 1999.* Retrieved from http://www.ehea.info/Uploads/Declarations/BOLOGNA_DECLARATION1.pdf

Evans, S. (2013). The long march to biliteracy and trilingualism: Language policy in Hong Kong education since the handover. *Annual Review of Applied Linguistics, 33*, 302–324. https://doi.org/10.1017/S0267190513000019

Evans, S., & Green, C. (2007). Why EAP is necessary: A survey of Hong Kong tertiary students. *Journal of English for Academic Purposes, 6*(1), 3–17. http://doi.org/10.1016/j.jeap.2006.11.005

Fang, F., & Liu, Y. (2020). 'Using all English is not always meaningful': Stakeholders' perspectives on the use of and attitudes towards translanguaging at a Chinese university. *Lingua, 247*, 102959.

Francomacaro, R. M. (2011). *English as a Medium of instruction at an Italian engineering faculty: an investigation of structural features and pragmatic functions.* [Doctoral dissertation, Università degli Studi di Napoli Federico II]. Retrieved from http://www.fedoa.unina.it/8637/1/francomacaro_mariarosaria_24.pdf.

Glenwright, P. D. L. (2003). Education reform and policy change in Hong Kong: A critique of the post-colonial legacy. *Education and Society, 21*(3), 67–89. Retrieved from http://repository.ied.edu.hk/dspace/handle/2260.2/6105

Graddol, D. (2006). *English next: Why global English may mean the end of 'English as a foreign language'.* Retrieved from www.britishcouncil.org/learning-research

Hamid, M. O., Nguyen, H. T. M., Baldauf Jr., R. B., & Baldauf, R. B. (2013). Medium of instruction in Asia: context, processes and outcomes. *Current Issues in Language Planning, 14*(1), 1–15. https://doi.org/10.1080/14664208.2013.792130

Hou, A. Y. C., Morse, R., Chiang, C.-L., & Chen, H.-J. (2013). Challenges to quality of English medium instruction degree programs in Taiwanese universities and the role of local accreditors: A perspective of non-English-speaking Asian country. *Asia Pacific Education Review, 14*(3), 359–370. https://doi.org/10.1007/s12564-013-9267-8

Hu, G., Li, L., & Lei, J. (2014). English-medium instruction at a Chinese university: Rhetoric and reality. *Language Policy, 13*(1), 21–40. https://doi.org/10.1007/s10993-013-9298-3

Hyland, K. (2003). *Second language writing.* Cambridge: Cambridge University Press.

Hyland, K. (2004). *Genre and second language writing.* Ann Arbor: Michigan University Press.

Hyland, K. (2007). Genre pedagogy: Language, literacy and L2 writing instruction. *Journal of Second Language Writing, 16*, 148–164.

Jäppinen, A. K. (2005). Thinking and content learning of mathematics and science as cognitional development in content and language integrated learning (CLIL): Teaching through a foreign language in Finland. *Language and Education, 19*(2), 147–168. https://doi.org/10.1080/09500780508668671

Kim, K. R. (2011). Korean professor and student perceptions of the efficacy of English-medium instruction. *Linguistic Research, 28*(3), 711–741.

Kirkpatrick, A. (2011). English as a medium of instruction in Asian education (from primary to tertiary): Implications for local languages and local scholarship. *Applied Linguistics Review, 2*, 99–120.

Kirkpatrick, A. (2014a). English as a medium of instruction in East and Southeast Asian universities. In N. Murray & A. Scarino (Eds.), *Dynamic ecologies a relational perspective on languages education in the Asia-Pacific region* (Vol. 9, pp. 15–24). Dordrecht: Springer. https://doi.org/10.1007/978-94-007-7972-3

Kirkpatrick, A. (2014b). The language (s) of HE: EMI and/or ELF and/or multilingualism? *The Asian Journal of Applied Linguistics, 1*(1), 4–15.

Kuteeva, M. (2020). Revisiting the 'E' in EMI: Students' perceptions of standard English, lingua franca and translingual practices. *International Journal of Bilingual Education and Bilingualism, 23*, 287–300.

Kwan, K. K. (1989). *An evaluation of students' language difficulties in using history and integrated science materials in form 1 in an Anglo-Chinese secondary school* [Master's thesis, University of Hong Kong]. Retrieved from http://hdl.handle.net/10722/51354

Lasagabaster, D., & Sierra, J. M. (2009). Immersion and CLIL in English: More differences than similarities. *ELT Journal, 64*(4), 367–375. https://doi.org/10.1093/elt/ccp082

Lavankura, P. (2013). Internationalizing higher education in Thailand: Government and university responses. *Journal of Studies in International Education, 17*(5), 663–676. https://doi.org/10.1177/1028315313478193

Legislative Council, L. (2009). *Fine-tuning the medium of instruction for secondary schools* (Vol. 09). Hong Kong. Retrieved from http://sc.legco.gov.hk/sc/www.legco.gov.hk/yr08-09/english/panels/ed/papers/ed0115cb2-623-1-e.pdf

Lei, J., & Hu, G. (2022). Research on English-medium instruction in the Asia-Pacific: Trends, foci, challenges, and strategies. In Lee, W. O., Brown, P., Goodwin, A. L., & Green, A. (Eds.), *International handbook on education development in Asia-Pacific* (pp. 1–23). Singapore: Springer.

Li, D. C. S. (2006). Chinese as a lingua franca in greater China. *Annual Review of Applied Linguistics, 26*, 149–176.

Lin, A. M. Y. (1997). Analyzing the 'language problem' discourses in Hong Kong: How official, academic, and media discourses construct and perpetuate dominant models of language, learning, and education. *Journal of Pragmatics, 28*(4), 427–440. https://doi.org/10.1016/S0378-2166(97)00031-3

Lin, A. M. Y. (2006). Beyond linguistic purism in language-in-education policy and practice: Exploring bilingual pedagogies in a Hong Kong science classroom. *Language and Education, 20*(4), 287–305. https://doi.org/10.2167/le643.0

Lin, A. M. Y., & Man, E. Y. F. (2011). The context and development of language policy and knowledge production in universities in Hong Kong. In K. Davis (Ed.), *Critical qualitative research in second language studies: Agency and advocacy on the Pacific Rim* (pp. 99–113). Greenwich: Information Age Publishing.

Lin, L. H. F., & Morrison, B. (2010). The impact of the medium of instruction in Hong Kong secondary schools on tertiary students' vocabulary. *Journal of English for Academic Purposes, 9*(4), 255–266. https://doi.org/10.1016/j.jeap.2010.09.002

Ljosland, R. (2011). English as an academic lingua franca: Language policies and multilingual practices in a Norwegian university. *Journal of Pragmatics, 43*(4), 991–1004. https://doi.org/10.1016/j.pragma.2010.08.007

Lo, Y. Y. (2014). L2 learning opportunities in different academic subjects in content-based instruction – evidence in favour of 'conventional wisdom.' *Language and Education, 28*(2), 141–160. https://doi.org/10.1080/09500782.2013.786086

Lo, Y. Y. (2015) How much L1 is too much? Teachers' language use in response to students' abilities and classroom interaction in Content and Language Integrated Learning. *International Journal of Bilingual Education and Bilingualism, 18*(3), 270–288.

Lo, Y. Y., & Macaro, E. (2012). The medium of instruction and classroom interaction: Evidence from Hong Kong secondary schools. *International Journal of Bilingual Education and Bilingualism, 15*(1), 29–52. https://doi.org/10.1080/13670050.2011.588307

Lo, Y. Y., & Macaro, E. (2015). Getting used to content and language integrated learning: What can classroom interaction reveal? *The Language Learning Journal, 43*(3), 239–255. https://doi.org/10.1080/09571736.2015.1053281

Long, M. H. (1983). Native speaker/non-native speaker conversation and the negotiation of comprehensible input. *Applied Linguistics, 4*(2), 126–141.

Macaro, E. (2018). *English medium instruction: Content and language in policy and practice*. Oxford: Oxford University Press.

Macaro, E. (2020). Exploring the role of language in English medium instruction. *International Journal of Bilingual Education and Bilingualism, 23*(3), 263–276.

Macaro, E., Curle, S., Pun, J., An, J., & Dearden, J. (2018). A systematic review of English medium instruction in higher education. *Language Teaching, 51*(1), 36–76. https://doi.org/10.1017/S0261444817000350

Mazak, C. M., & Herbas-Donoso, C. (2015). Translanguaging practices at a bilingual university: A case study of a science classroom. *International Journal of Bilingual Education and Bilingualism, 18*, 698–714.

Ministry of Education Malaysia. (2013). *Malaysia Education Blueprint 2013–2015 (Preschool to Post-Secondary Education)*. Retrieved from https://www.moe.gov.my/menumedia/media-cetak/penerbitan/dasar/1207-malaysia-education-blueprint-2013-2025/file

Othman, J., & Mohd Saat, R. (2009). Challenges of using English as a medium of instruction: Pre-service science teachers' perspective. *The Asia-Pacific Education Researcher, 18*(2), 307–316.

Pakir, A. (2004). Medium-of-instruction policy in Singapore. In J.W. Tollefson & A. B. M. Tsui (Eds.), *Medium of instruction policies: Which agenda? Whose agenda?* (pp. 117–134). London: Routledge.

Pecorari, D., & Malmström, H. (2018). At the crossroads of TESOL and English medium instruction. *Tesol Quarterly, 52*(3), 497–515.

Poon, A. Y. K. (2004). Language policy of Hong Kong: Its impact on language education and language use in post-handover Hong Kong. *Journal of Taiwan Normal University, 49*(1), 53–74.

Poon, A. Y. K. (2009). A review of research in English language education in Hong Kong in the past 25 years: Reflections and the way forward. *Educational Research Journal, 24*(1), 7–40.

Poon, A. Y. K. (2010). Language use, and language policy and planning in Hong Kong. *Current Issues in Language Planning, 11*(1), 1–66. https://doi.org/10.1080/14664201003682327

Poon, A. Y. K. (2013). Will the new fine-tuning medium-of-instruction policy alleviate the threats of dominance of English-medium instruction in Hong Kong? *Current Issues in Language Planning, 14*(1), 34–51. https://doi.org/10.1080/14664208.2013.791223

Poon, A. Y. K., Lau, C. M. Y., & Chu, D. H. W. (2013). Impact of the fine-tuning medium-of-instruction policy on learning: Some preliminary findings. *Literacy Information and Computer Education Journal (LICEJ), 4*(1), 946–954.

Probyn, M. (2001). Teachers voices: Teachers reflections on learning and teaching through the medium of English as an additional language in South Africa. *International Journal of Bilingual Education and Bilingualism, 4*(4), 249–266. https://doi.org/10.1080/13670050108667731.

Probyn, M. (2006). Language and learning science in South Africa. *Language and Education, 20*(5), 391–414.

Probyn, M. (2009). 'Smuggling the vernacular into the classroom': Conflicts and tensions in classroom codeswitching in township/rural schools in South Africa. *International Journal of Bilingual Education and Bilingualism, 12*(2), 123–136. https://doi.org/10.1080/13670050802153137

Pun, J., & Macaro, E. (2018). The effect of first and second language use on question types in English medium instruction science classrooms in Hong Kong. *International Journal of Bilingual Education and Bilingualism, 22*(1), 1–14.

Pun, J. K. H., & Thomas, N. (2020). English medium instruction: teachers' challenges and coping strategies. *ELT Journal, 74*(3), 247–257. https://doi.org/10.1093/elt/ccaa024

Rose, H., & Galloway, N. (2019). *Global Englishes for language teaching.* Cambridge: Cambridge University Press.

Rose, H., McKinley, J., Xu, X., & Zhou, S. (2020). *Investigating policy and implementation of English medium instruction in higher education institutions in China.* London: British Council.

Sahan, K., Galloway, N., & McKinley, J. (2022). 'English-only' English medium instruction: Mixed views in Thai and Vietnamese higher education. *Language Teaching Research*, 136216882110726. https://doi.org/10.1177/13621688211072632

Sahan, K., & Rose, H. (2021). Translanguaging or code-switching?: Re-examining the functions of language in EMI classrooms. In Di Sabato B., Hughes B. (Eds.), *Multilingual perspectives from Europe and beyond on language policy and practice* (pp. 45–62). London: Routledge.

Schleppegrell, M. J. (2004). *The language of schooling: A functional linguistic perspective.* Mahwah, N.J.: Lawrence Erlbaum.

Snow, D. (2004). *Cantonese as written language: The growth of a written Chinese vernacular.* Hong Kong: Hong Kong University Press.

Snow, M. A., & Brinton, D. M. (2023). *Content-based instruction: What every ESL teacher needs to know.* Michigan: University of Michigan Press.

Standing Committee on Language Education and Research. (2003). *Action plan to raise language standards in Hong Kong: Final report of language education review.* Retrieved from http://www.language-education.com/eng/ publications_actionplan.asp

Studer, P. (2014). Coping with English: Students' perceptions of their teachers' linguistic competence in undergraduate science teaching. *International Journal of Applied Linguistics, 25*(2), 183–201. https://doi.org/10.1111/ijal.12062

Swain, M., & Johnson, R. K. (1997). Immersion education: A category within bilingual education. In R. K. Johnson & M. Swain (Eds.), *Immersion education: international perspectives* (pp. 1–16). Cambridge: Cambridge University Press.

Swain, M., & Lapkin, S. (2005). The evolving sociopolitical context of immersion education in Canada: Some implications for program development. *International Journal of Applied Linguistics, 15*(2), 169–186. https://doi.org/10.1111/j.1473-4192.2005.00086.x

Tamtam, A. G., Gallagher, F., Olabi, A. G., & Naher, S. (2012). A comparative study of the implementation of EMI in Europe, Asia and Africa. *Procedia - Social and Behavioral Sciences, 47*, 1417–1425. https://doi.org/10.1016/j.sbspro.2012.06.836

Tatzl, D. (2011). English-medium master's programmes at an Austrian university of applied sciences: Attitudes, experiences and challenges. *Journal of English for Academic Purposes, 10*(4), 252–270. https://doi.org/10.1016/j.jeap.2011.08.003

The Education University of Hong Kong. (2021). *The University's Language Policy.* Retrieved from https://www.eduhk.hk/acadprog/downloads/Language%20Exit%20Requirements_UGC.pdf

Thøgersen, J., & Airey, J. (2011). Lecturing undergraduate science in Danish and in English: A comparison of speaking rate and rhetorical style. *English for Specific Purposes, 30*(3), 209–221. https://doi.org/10.1016/j.esp.2011.01.002

Tsui, A. B. M. (2004). Medium of instruction in Hong Kong: One country, two systems, whose language? In J.W. Tollefson & A. B. M. Tsui (Eds.), *Medium of instruction policies: Which agenda? Whose agenda?* (pp. 97–116). London: Routledge.

Tsui, A. B. M. (2007). Language policy and the construction of identity: The case of Hong Kong. In A.B.M. Tsui, & J.W. Tollefson (Eds.), *Language policy, culture, and identity in Asian context* (pp. 121–141). London: Routledge.

Van Deusen-Scholl, N., & Hornberger, N. (2008). *Second and foreign language education: Encyclopedia of language and education.* New York: Springer.

Wang, W., & Curdt-Christiansen, X. L. (2018). Translanguaging in a Chinese–English bilingual education programme: A university-classroom ethnography. *International Journal of Bilingual Education and Bilingualism, 22*(3), 322–337. https://doi.org/10.1080/1367 0050.2018.1526254

Wannagat, U. (2007). Learning through L2 – Content and Language Integrated Learning (CLIL) and English as Medium of Instruction (EMI). *International Journal of Bilingual Education and Bilingualism, 10*(5), 663–682. https://doi.org/10.2167/beb465.0

Yip, D. Y. (2004). Questioning skills for conceptual change in science instruction. *Journal of Biological Education, 38*(2), 76–83. https://doi.org/10.1080/00219266.2004.9655905

Yip, D. Y., Coyle, D., & Tsang, W. K. (2007). Medium of instruction on science students: Instructional activities in science lessons. *Education Journal, 35*(2), 77–107.

Yip, D. Y., Tsang, W. K., & Cheung, S. P. (2003). Evaluation of the effects of medium of instruction on the science learning of Hong Kong secondary students: Performance on the science achievement test. *Bilingual Research Journal, 27*(2), 295–331. https://doi.org/10 .1080/15235882.2003.10162808

Zhao, T., & Macaro, E. (2014). What works better for the learning of concrete and abstract words: Teachers' L1 use or L2-only explanations? *International Journal of Applied Linguistics, 26*(1), 75–98. https://doi.org/10.1111/ijal.12080

2 EMI policies and practices

2.1 Classroom interactions in EMI classrooms

It has been argued that classroom interactions will not be rich enough for 'deep' learning unless students have a high level of English proficiency. Research on classroom interactions in the context of English medium instruction (EMI) has focused on teacher–student interactions and teaching practices of individual science teachers in EMI classrooms (Chan, 2013, 2014; Lin, 2006; Lo & Macaro, 2012; Lo, 2014; Wannagat, 2007; Yip et al., 2007a, b). These studies employed ethnographic approaches (e.g., observations and interviews) together with discourse analysis to identify patterns of interactions between teachers and students. By using mixed-methods designs when studying EMI classrooms, researchers can understand complex psychological constructs such as students' beliefs about classroom interactions (An & Thomas, 2021; Bernat & Gvozdenko, 2005; Sato, 2013). These studies describe different types of talk and pedagogies for delivering scientific content through English or code-mixing between the L1 and English.

Wannagat (2007) observed and video-recorded the classroom interactions in EMI and CLIL settings from two English-taught secondary Grade 9 and 10 history classes in Hong Kong and Germany. He found that EMI students in Hong Kong had more exposure to English, with over 50% of classroom discourse conducted in English, whereas CLIL students in Germany had only 20% of two or three subjects conducted in English. This suggests that EMI seems to be more successful than CLIL. He added that although EMI teachers seemed to take the language issue into account by using simplified texts and providing key L1 terms, the role of language in learning objectives was not clearly explicated in the EMI policy. As such, the language issue is often largely ignored in curriculum development and teacher training. Thus, in Wannagat's view, EMI is less successful than CLIL in reality, given the absence of language focus in the EMI curriculum.

Lo and Macaro (2012) used classroom observations to explore classroom interaction between EMI and MOI-switching secondary schools (e.g., schools that switched from Chinese MOI to allowing EMI in secondary Grade 9) in Hong Kong. They compared the quality of L2 language learning in classrooms for

DOI: 10.4324/9781003001454-3

humanities and science subjects by the number of student turns, length of student turns and initiation–response–feedback (IRF) patterns (Sinclair & Coulthard, 1975). The study also compared types of focus-on-form exchanges by paying attention to specific grammatical rules and lexical items (see Costa, 2012). After observing 22 lessons from Grades 9 and 10, they found that science lessons were teacher-centred, teacher–student interactions were reduced to IRF sequences, the proportion of student talk was significantly lower, and student turns tended to be short. Teachers also paid less attention to form-focused instruction in the L2. The register of science and the availability of other semiotic resources for teaching science largely reduced opportunities for students to engage in L2 language learning through discussions and to interact with their teacher. For example, teachers would draw a diagram to explain the process of photosynthesis and consequently spend less time explaining the concepts to students with spoken interactions because they believed that students would understand by looking at the diagram.

Lo and Macaro (2015) conducted another study to investigate the differences in overall interaction patterns between late and very late EMI classrooms with students at the same academic level (Grade 10). 'Late' and 'very late' refer to schools that switch to English instruction at junior secondary (Grades 7–9) and senior secondary (Grades 10–12), respectively. They collected 15 audio/video-recorded classroom observations of Grade 10 content subjects (e.g., history, geography, biology, physics) with eight content subject teachers and approximately 320 students from one 'very late' MOI-switching and two full EMI schools in Hong Kong. The observed interactive patterns were analysed quantitatively in terms of the distribution and functions of teacher and student talk, duration of teacher and student turns, IRF exchange sequence patterns between teacher and student, and teachers' questioning skills.

Lo and Macaro (2015) found that students in 'very late' MOI-switching schools, who had limited EMI experience, lacked fluency in their L2 utterances. This was manifested in their longer turns with less content richness (measured in syllables). When students interacted with their teachers, there were fewer verbal exchanges observed, and these exchanges were usually used in much simpler forms of IRF sequences (e.g., yes or no responses). This suggests that the teachers were less able to modify their questions to elicit additional responses after students failed to give the expected answer to the original question. The study concluded that the teaching practices in this MOI-switching school were teacher-centred and monologic. Students had difficulties communicating their ideas in the L2 and the question–answer sequences were constrained. Based on these findings, Lo and Macaro recommended that teachers and students should be allowed to use code-switching during the early stages of EMI adoption to prepare for full immersion in the L2. Through code-switching (translation), teachers can help students understand content accurately and solve any language problems more easily (Hong, 2022).

According to Cummins (1979), it might take at least three to five years for students to successfully gain the proficiency needed to benefit fully from immersion programmes. Students also need to attain a threshold level of English proficiency to allow for effective spoken interactions in classrooms. Further, teachers need to be familiar with EMI instruction strategies and develop the necessary teaching skills to interact with students in extended verbal exchanges.

Other studies have reported that science teachers have to deal with diverse levels of student English proficiency (Probyn, 2001, 2006, 2009). As noted by Soruç et al. (2021), one of the biggest challenges is led by the diversity of the linguistic landscape in the EMI classroom. This is due to the fact that the teaching of the English language is neither prioritised nor well-supported (Soruç et al., 2021). As a result, some students' language proficiency is inadequate to comprehend the academic content in the curriculum (Soruç et al., 2021). These students may encounter difficulties across all four language skills, namely speaking, listening, reading, and writing (Soruç et al., 2021). Some of these language barriers have been identified by certain studies in different cultural contexts such as Hong Kong (Evans & Morrison, 2011). Substantial research in Hong Kong on learning science through English indicates that students have inadequate English proficiency (Marsh et al., 2000; Yip et al., 2007a, b). Some studies have compared the English achievement of humanities and science students in terms of the EMI classroom; for example, Lo (2014) argued that there are more English learning opportunities in humanities EMI classrooms because teachers and students interact more frequently and that students have many chances of producing additional language in these classrooms. However, none of these studies offers empirical evidence that shows that language learning will be automatically facilitated once students study humanities subjects using the English language, nor that they are unable to learn the language effectively during science classes (Lo, 2014). Therefore, limited research has indicated whether the English achievement of science students in Hong Kong is poorer than that of students in humanities subjects. This is difficult to investigate as, under the 3-3-4 education curriculum implemented in 2009, students are no longer identified as either arts- or science-focused.

Lin (2006) examined classroom interactions in the 1980s and found that in some of the bilingual classrooms with effective teachers identified by the school principals and the Education Bureau, interactions promoted higher cognitive gain and conceptual change in students. She concluded that practical bilingual pedagogies could be drawn from these classrooms through the use of students' available linguistic resources, including rich L1 semantic context, code-mixing strategies, and specific IRF patterns in L1 followed by a recap in L2. However, a study by Jiang et al. (2016) examined how EMI instruction is implemented at the tertiary level and explored the teachers' perceptions and students' motivations in EMI. Drawing on observations of nine classrooms, three post-observation interviews, and one questionnaire, the authors point out that although code-switching may facilitate students' content understanding, the meaning construction and transmission in L2 depended primarily on written texts on the PowerPoint slides. As a result, this code-mixing instruction perhaps only developed learners' L1 and L2 receptive skills (reading comprehension and listening), rather than developing their productive skills in L2 (writing and speaking). The results of code-mixing instruction could suggest that many of these observed EMI classrooms have provided very few opportunities to develop students' disciplinary English skills in a comprehensive manner.

Lin and Wu (2015) conducted a detailed analysis of five minutes of classroom discourse from an EMI junior secondary science classroom in Hong Kong with below-average students. The researchers provided an example of how a skilled bilingual science teacher interacts with the students through their everyday language in L1 in order to develop their disciplinary language in L2. In their analysis, Lin and Wu show that the teacher first engages with students through 'languaging' (Swain, 2009), which serves to mediate second language learning, a vehicle through which thinking is articulated and transformed into an artefactual form. In second language acquisition (SLA) classroom contexts, languaging is a pedagogical stance that teachers and students take on that allows them to draw on all of their linguistic (e.g., L1, L2) and semiotic (e.g., symbols) resources as they teach and learn both language and content materials (Mazak, 2016). This languaging process allows students to fully understand the content by operating between L1 and L2.

This is achieved by a number of teacher-led questions in the form of IRF patterns, designed to help students master the thematic patterns of scientific discourse. This study demonstrates that in the Hong Kong context, even students with low English proficiency are able to develop their disciplinary language in L2 in the content subject classroom. However, this can only happen if the bilingual teacher provides sufficient teacher-led guidance and comprehensible input in the form of languaging between L1 and L2. Under these conditions, students are able to understand the content knowledge and thus contribute to class discussion in L1, while gradually developing knowledge of the equivalent meanings in L2. Lin and Wu argue that the role of a content subject teacher is as a bilingual teacher, and they demonstrate the skill of these teachers in asking meaningful questions that can create space for translanguaging, thus developing content knowledge. Pun and Tai (2021) have also proven that EMI science laboratory sessions can function as a translanguaging space that offers an interactional space for teachers and students to make use of multilingual and multimodal resources in order to learn science knowledge and develop their ability to perform scientific tasks.

In many observational studies, researchers of science education have proposed that skilful questioning by teachers can lead to meaningful learning in science, which requires interactive teaching styles that encourage students to express their views and ask questions, and that promote a learning culture based on discussion and other student-centred activities (Yip et al., 2003, 2007; Yip & Tsang, 2006; Yip, 1999, 2001, 2004). A study by Yip et al. (2003) found that Hong Kong science teachers who were identified as 'generally skilful' at questioning tended to use a wide range of question types in their classrooms (e.g., lower-order, higher-order, motivational, and concept-changing). However, while the teachers aimed to use these questions in order to stimulate higher-order cognitive thinking, Yip pointed out that such aims were often not achieved because the teachers did not make use of students' prior experiences and existing knowledge. Yip's study indicates that science teachers tend to use recall questions with lower cognitive demands for a number of reasons: (1) the time constraints on teaching, (2) the pressure to cover the examination syllabus within a tight schedule (Yip, 1999), (3) the low English proficiency of students, and (4) the poor communication skills of some EMI science teachers (Yip & Tsang, 2006). Some of these science teachers do not have adequate pedagogical content knowledge, due to the fact that their teaching duties do not match their science qualifications. Thus, many of them have to teach outside their subject specialisations, which greatly reduces the quality of teaching and the teaching strategies from which they can draw (Yip, 2003).

In a later study, Yip (2004) observed Hong Kong science teachers' questioning techniques. He identified ten types of teacher questions that induced conceptual changes in their science students' learning behaviours. These question types are lower-order (recalling, explanation), higher-order (analysis, evaluation, synthesis), motivation and conceptual change (eliciting, challenging, extending, application). Similar to the previous study, Yip's results showed that science teachers in Hong Kong are supposedly skilful at questioning due to asking a relatively high number of higher-order questions in their classrooms. However, although some teachers use higher-order questions to promote students' higher-order thinking, they rarely consider students' prior knowledge when asking these questions. As such, the pedagogical values of these questions are not achieved, and this leads to student confusion and frustration as they lack understanding of the concepts required to answer these questions. Faced with a lack of response from their students, most teachers simply gave the correct answers or focused on questions with a lower cognitive demand. The study also found that some teachers use a number of conceptual questions to identify student misconceptions, but the teachers fail to promote students' cognitive thinking in resolving inconsistent ideas and constructing new ideas from their existing knowledge. Yip's study shows that there is a relationship between the types of questions used by teachers and their pedagogical functions in students' learning in science. His study provides a valuable framework for researchers to examine teacher–student interaction in a science classroom. For instance, skilful teachers use more 'conceptual change' questions to elicit preconceptions and misconceptions from their students and promote 'meaningful learning' in science (Yip, 2004, p. 79).

The research mentioned above shows that EMI science classrooms in Hong Kong generally take a teacher-centred approach. These practices can be regarded as a strategy to reduce the linguistic demands of the lesson and cater to students' limited English proficiency. However, by lowering their expectations or 'downgrading' their teaching practices, teachers may have a substantial negative effect on the quality of the learning experience in a science classroom (Yip, 1999). For example, students have little opportunity to practice how to present abstract and complex concepts or to express their ideas or arguments systematically in English during class. A reduced quality of teaching in science indirectly promotes rote learning (Poon, 2013) and makes learning more of a mechanical process (Yip & Tsang, 2006).

It is also important to consider teachers' beliefs about EMI as their perceptions could have an impact on their teaching practices. As Tung et al. (1997) point out, decisions about whether teachers adopt full EMI, full CMI, or use code-mixing between Cantonese and English are influenced by the realities of the classroom (e.g., students' English proficiency), and whether their students can achieve the school's educational goals. The next sections will describe previous studies that examined teachers' and students' perceptions of EMI experiences.

2.2   Teachers' attitudes towards EMI teaching and their teaching practices in EMI classrooms

One of the underlying issues in Hong Kong's MOI debates (and debates on the adoption of EMI to teach elsewhere) is whether obliging ESL students to study a subject area in a full English immersion context would result in lower quality of learning the subject matter compared with instruction in the L1, regardless of the students' potential gains in English proficiency. Many studies have reported that teachers face numerous issues when teaching solely through English in their classrooms (Dearden, 2015; Llurda, 2013; Tatzl, 2011). Othman and Mohd Saat (2009) reported that pre-service science teachers in Malaysia encounter challenges even though they receive some EMI training. The challenges include a lack of teaching skills for integrating both content and language teaching, no appropriate instructional materials for teaching both content and language and poor English ability. Teachers in many countries receive no specific EMI training (Dearden, 2015), with the result that many teachers do not develop the necessary linguistic competence and pedagogical skills to deliver content knowledge in English.

In addition, teachers generally resist being made responsible for teaching both content and language (Probyn, 2001). Unlike European CLIL teachers who accept the emphasis on learning both content and language in their teaching curriculum (Coyle et al., 2010), most EMI teachers around the world believe that their responsibility is only to teach content knowledge but not language (Airey, 2012). Probyn (2006) interviewed EMI science teachers in Africa, a number of whom claimed that teaching English in science was the responsibility of English language teachers. The science teachers in Probyn's study said they should not be expected to make science content more accessible to their students in English. Probyn's studies have shown that many EMI science teachers might not believe they have a role in teaching the language of science in addition to content.

In Hong Kong, teachers have exhibited similar attitudes in EMI science classrooms. In a study of Hong Kong junior secondary science teachers, Hoare (2003) found that many EMI science teachers did not have very developed 'language awareness', which refers to an enhanced consciousness and sensitivity towards the forms and functions of language. This has led them to pay less attention to their students' L2 language needs in the EMI classroom. Hoare's study recognised the fact that teaching a content subject is more complicated than translating what one knows into English.

Yip and Tsang (2007) examined the impact of EMI on science teaching and the learning process in junior secondary EMI classrooms. Their observations indicated that the participating science teachers were mostly unprepared for integrating science and language. These teachers lacked the communicative competence to explain abstract and complex scientific concepts in English and thus were reluctant to use higher-order and conceptual change questions for science teaching. The observed EMI science teachers had poor communication skills and thus may have lacked confidence in teaching science through English (Yip et al., 2007).

Lo and Macaro (2012) compared teacher–student interactions in both junior science and humanities classrooms in Hong Kong. They observed a very minimal frequency of teacher–student interaction and short IRF sequences in the science classrooms. EMI teachers focused only on lower-order questions, by asking students to recall factual information in science. Teachers rarely provided corrective feedback on students' L2 language production (Lo, 2014).

According to Mortimer and Scott's science communication framework (2003), the teaching practices observed in previous studies (Lo & Macaro, 2012; Yip, 2003, 2004) employ a non-interactive and authoritative communicative approach. Such a teacher-centred teaching style is likely to greatly reduce the opportunity for meaningful interactions, leading to L2 production between teachers and students (Yip et al., 2003). In science education research, teacher and student interactions are important for students to develop the scientific enquiry skills that are essential to science learning (Wellington & Osborne, 2001). According to Long's (1983) interaction hypothesis in the field of SLA, meaningful interaction helps to promote students' natural L2 development. Students acquire the necessary L2 linguistic competence through comprehensible input and guided feedback from their teachers (Krashen, 1982). If teachers do not make efforts to interact with students, their students' learning may be negatively affected.

This is evident in Yip et al.'s (2007) study, which observed the quality of instructional activities in Hong Kong EMI science classrooms and showed that many teachers adopt a monologic teaching style, which causes the quality of instructional activities to suffer. They suggested that a possible reason for the monologic teaching style is teachers' perceptions of EMI. In these monologic EMI classrooms, students spent most of their time listening to teachers' explanations, watching demonstrations and following teachers' manuals when performing experiments. The possible consequences of such a monologic teaching style on the students' learning are well documented in the literature, as the following summary indicates.

Firstly, students have fewer opportunities for knowledge construction and meaning negotiation, which is among the key pedagogies that can help students construct an identity as novice members of a scientific community of practice (Evnitskaya & Morton, 2011). This factor will be discussed further in Section 2.3. Secondly, there are fewer interactions between teachers and students, despite the fact that these interactions can be valuable opportunities for teachers to evaluate students' understanding, correct their inconsistent concepts, identify misconceptions, and promote conceptual change from their existing knowledge (Yip, 2004). Thirdly, when teaching follows a teacher-centred approach with emphasis on memorisation and instruction, students are engaged only in rote learning (Poon, 2013) and therefore have relatively fewer opportunities to design their own methods of investigation, which would require greater creativity and higher-order thinking skills (Yip & Tsang, 2006).

In sum, the literature clearly documents the teaching challenges that most teachers experience in EMI classrooms. These challenges cannot be easily resolved, even by teachers who have received EMI training. These teachers have shaped their science teaching style to be monologic, teacher-centred, and based on rote learning, due to the belief that they are not responsible for teaching the language of science. Such a teaching style may have a negative impact on the quality of instructional activities in the classroom, on the interactions with students and on the achievement of learning goals. However, the extent of this negative impact depends on the coping strategies these EMI teachers use to overcome the challenges that can arise from this teaching style, such as fewer opportunities for knowledge construction and meaning negotiation, difficulties articulating ideas in the L2, and constrained question–answer sequences. Researchers have indicated the importance of helping teachers develop complete understandings and "can do" attitudes to increase the frequency of the use of bi/multilingual strategies relevant to their pedagogical goals (Mandoza & Ou, 2022). Therefore, the following section describes some of the coping strategies reported in previous studies that EMI teachers use to resolve teaching challenges in EMI classrooms.

2.3   Coping strategies related to the teaching process

In order to handle the duties of teaching both scientific content and language, Yu (2002) suggests that teachers can provide comprehensible input to integrate language and science instruction through a range of instructional activities. These activities could include (1) promoting language use and scientific thinking through group work, (2) explicitly teaching learning skills, (3) acknowledging cultural differences, (4) using modified classroom talk, (5) appreciating diversity, (6) using students' prior knowledge in science, (7) enriching the curriculum, and (8) providing a rich L2 language use environment.

Researchers have also observed that effective science teachers use a range of semiotic supports to deliver science content in English. Flowcharts, diagrams, body language, visual aids, practical demonstrations and multi-media software can help students make connections between the content and their everyday experiences (Ferreira, 2011). Noting that students' ability to understand examination questions is directly related to their science achievement, researchers recommend that teachers use simple English with short phrases on exam papers, thus avoiding student misinterpretation of exam questions and allowing for more accurate measurement of their science achievement (Prophet & Badede, 2007).

Other researchers suggest strategies to make the disciplinary language in L2 more accessible to students by (1) using blackboards for illustrations; (2) encouraging students to speak English by extending their vocabulary (Probyn, 2006); (3) using translation, simplification and emphasis on keywords; and (4) speaking slowly to ensure that students fully understand the content (Huang & Normandia, 2008; Tan, 2011). Othman and Mohd Saat (2009) stress that it is important for EMI teachers to introduce new vocabulary before teaching a topic to ensure students' full understanding.

Strategies suggested to increase the quality and frequency of classroom interactions in different language ability groups include code-switching with students with poor English proficiency, along with extended turn exchange sequences of IRFRF (Initiation–Response–Feedback–Response–Feedback) (Lo & Macaro, 2012). Lin (2006) analysed teaching practices in Hong Kong science classrooms and proposed that a practical bilingual pedagogical approach, where most subject content is delivered in L1 (except for key terms and recapping in L2), would allow students more time to make the transition from L1 to L2. This approach could help junior secondary CMI science students with poor English proficiency to access more technical terms in English and understand the language style of scientific discourse by using their L1 and cultural resources. Teachers interviewed about their EMI classroom teaching across various studies (Ferreira, 2011; Yip et al., 2007) have reported using different coping strategies to help students respond to questions in English, express abstract concepts, understand scientific terminology, construct scientific knowledge with their peers and help manage student discipline and interpersonal relations.

Evnitskaya and Morton (2011) conducted a conversation analysis of two secondary CLIL biology classrooms in Spain, a context similar to EMI with a focus on both content and language learning. They found teachers and students interact in ways consistent with membership of a 'community of practice' (CoP) (Wenger, 1998). CoP is a concept that refers to members in a community sharing common goals and interests and pursuing these goals through engaging in joint activities. The teachers and students observed in the Spanish biology classrooms interacted to share both linguistic and other non-verbal meaning-making resources, developed joint learning goals and engaged as a group when doing school science. A highly interactive classroom is believed to improve student autonomy in undertaking scientific enquiry and to encourage students to feel membership in a scientific CoP through the co-construction of knowledge with teachers and their peers.

Some science teachers have attempted to foster students' cognitive development through higher-order questions. Yip (2004) explored the impact of training science teachers with a focus on a conceptual change model of science instruction. The conceptual change suggests that teachers correct students' misconceptions about science through questioning. The study observed the classrooms of 14 in-service science teachers with three to six years of teaching experience who were taking the final year of the Chinese University of Hong Kong's Postgraduate

Diploma in Education (PGDE) Programme. Two assessors reviewed 405 questions asked by the teachers and categorised them according to their pedagogical functions. Yip identified ten question types: lower-order (recalling, explanation), higher-order (analysis, evaluation, synthesis), motivation and conceptual change (eliciting, challenging, extending, application). These ten question types served four different pedagogical functions that were said to encourage conceptual changes in students' understandings of science: (1) check students' understanding of factual information and their ability to explain a science phenomenon (lower-order); (2) evaluate their higher cognitive skills (higher-order); (3) investigate students' learning motivation; and (4) lead to conceptual changes in a student's science knowledge.

Yip suggests a range of questioning techniques that effective science teachers use in Hong Kong classrooms to evaluate students' understanding, correct inconsistent concepts, identify misconceptions and promote conceptual change from students' existing knowledge. Similarly, Chin (2007) observed 36 Grade 7 science classrooms in Singapore and identified four types of questioning approaches used by science teachers to scaffold students' thinking and construct scientific knowledge: Socratic questioning, verbal jigsaw, semantic tapestry, and framing. The following paragraph defines these questioning approaches.

As described by Chin (2007), Socratic questioning is an approach based on thoughtful dialogue, wherein teachers claim ignorance of the topic while using questions to prompt and guide student thinking. This enables students to examine ideas based on logic and prior knowledge while promoting independent thought and higher-level thinking skills. Verbal jigsaws involve the use of topic-specific terms to form integrated proposition statements. In the case of science classrooms, students are prompted to create associations between scientific terms to build a coherent mental framework. It is often used with topics that contain several technical terms or for students with weaker language skills. Semantic tapestry is an approach that focuses on ideas and abstract concepts. It helps students connect different ideas together into a conceptual framework by posing questions that address multiple aspects of a problem and stimulates multimodal thinking through talk, diagrams, visual images, symbols, formulas, and calculations. Finally, framing uses questions to contextualise a topic and structure the related discussion. By employing this approach, questions can act as precursors to subsequent information, broad outlines for the discussion, or question–answer summaries of key points. This helps students with understanding the relationship between a question and the information it addresses.

Othman and Mohd Saat (2009) conducted a questionnaire survey to explore the instructional strategies that pre-service teachers use in Malaysia's EMI science classrooms when overcoming the language-related challenge of delivering science content in English. They report that effective pre-service science teachers use questions to help students construct concepts, use higher-order questions to induce cognitive thinking, use probes to explore preconceptions and create constructive chalkboard layouts in a logical sequence to summarise learned concepts. To deal with different ability groups, Ferreira (2011) conducted focus group interviews with 28 life sciences teachers in South Africa with between 16 and 29 years of teaching experience to understand the coping strategies they used to help their English second language learners. He points out that teachers can group learners with the same level of science achievement but with different language competence in classroom activities, creating opportunities for those with greater understanding to help their peers.

Science teacher training programmes have been found not to equip teachers with the necessary linguistic competence and pedagogical skills to teach science in English in many EMI contexts such as Hong Kong and Malaysia (Othman & Mohd Saat, 2009; Yip, 2004; Yip et al., 2003; Walker, 2010). For example, although pre-service teachers in Malaysia receive some training in teaching science, studies have identified a lack of teaching skills in integrating content with language teaching and a lack of appropriate instructional materials (Mohd Saat & Othman, 2010; Othman & Mohd Saat, 2009). Yip et al. (2007a, b) believe that science teachers need to develop and use a repertoire of context-embedded language skills (e.g., everyday English) to convey meaning.

Walker's study is echoed by researchers in functional linguistics, who see language and learning as social processes. They believe that language use is shaped by the kind of activity taking place and who is doing it (Halliday & Matthiessen, 2004). Language structure is an integral part of a text's social context and function (Martin & Rose, 2008). These functional linguists believe that an understanding of registers and genres in scientific texts significantly affects a learner's ability to read and write effectively in the science community (Halliday & Martin, 1993; Martin & Veel, 1998; Schleppegrell, 2004). The European CLIL context has now introduced a genre-based pedagogy into content subject classrooms as an attempt to address the academic literacy challenges faced when students engage with different text types. The intent is to improve their language competence in producing subject-specific discourse with appropriate choices of vocabulary, grammar, and language style (Llinares et al., 2012; Lorenzo, 2013).

In addition, teachers' language awareness has been shown to be closely associated with the effectiveness of the language learning process for students (Hoare, 2003). Responsibility for language awareness is also relevant for those who teach content subjects through English. Language awareness can be realised through teaching practices and operationalised through self-designed teaching interventions, such as using simple English to make science more accessible to students (Hoare, 2003). Hoare (2003) conducted an empirical study to investigate any differences in teaching practices between teachers with high language awareness (e.g., those who pay attention to students' English needs in learning science) and those with low language awareness in Hong Kong junior secondary science classrooms. He examined six junior secondary science teachers in EMI classrooms with multiple sources of data, including interviews with teachers and students, a teachers' questionnaire about their degree of language awareness, and a test of their students' science achievements. The findings show that students taught by teachers with higher language awareness are better able to express their understanding of science learned through English. There is also a qualitative difference in teaching practices. Teachers with higher language awareness were observed to operationalise their awareness as strategies to make scientific meanings more explicit and accessible to their students through English. Other researchers (Arkoudis, 2005; Barwell, 2005; Lo, 2014; Poon, 2013) recommend collaborative teaching with English language teachers who can inform science teachers about the language features necessary for science learning in an English environment.

Literature on teachers' professional development suggests that science teachers should be given more opportunities to review and strengthen their subject knowledge, skills, and confidence in teaching for conceptual change (Mak et al., 1999; Yip, 2001). Recommendations include mixed discipline peer reviews in which groups of science teachers with different specialisations reflect on their own preconceptions. Some, such as Yip (1999), suggest that a cultural change is needed to create more supportive cultures within schools and a collaborative team across different science departments. For example, senior science teachers in each school could offer school-based advice and experience on subject matter development (CDC & HKEAA, 2007). A collaborative team with teachers from all science departments in a school, together with internal and external support for staff development, can help introduce and implement innovations, such as new teaching interventions for conceptual change in the classroom (Lo, 2014; Yip, 1999).

In sum, the previous studies have documented some of the strategies content teachers can use to attempt to resolve the teaching challenges of the EMI context. These strategies include:

- Combining content and language learning in different instructional activities.
- Making L2 disciplinary language more accessible verbally and non-verbally, by drawing from students' everyday experiences.
- Improving the quality of teacher–student interaction through extensive IRF patterns.
- Using the L1 effectively in scaffolding.
- Improving teachers' questioning techniques.
- Increasing students' understanding of academic registers and genres.
- Increasing teacher's language awareness, providing more professional development for content teachers to strengthen their subject knowledge.
- Developing collaborative teaching with language teachers.

While teacher–student interactions are reciprocal, it is worth noting some of the students' experiences in EMI, their achievements, their challenges, and their coping strategies in EMI classrooms. The following section discusses studies that have examined students' perceptions of their EMI experience. The specific focus is on the two areas previously examined in MOI studies in Hong Kong that relate to students' EMI experience: (1) students' achievements in academic subjects and (2) students' self-concepts in science (Lo & Lo, 2014; Marsh et al., 2000; Yip & Tsang, 2007a, b).

2.4 Students' views on their EMI experiences, achievements, and self-concepts

2.4.1 Research on EMI students' achievements in academic subjects

Research into language policy provides empirical evidence of the correlation between the impact of different MOIs (Chinese and English) and students' academic achievement. Academic achievement refers to a student's school examination results in his or her language subjects (English and Chinese) and content subjects (like mathematics and science). Two studies have explored the scientific achievements of Hong Kong students at junior and senior secondary levels in an English learning environment (Marsh et al., 2000; Yip & Tsang, 2006).

At junior secondary level, Yip and Tsang (2006) conducted a longitudinal study with a sample of 12,000 students to investigate the effects of the compulsory mother-tongue policy in 1998 on science learning during the first two years of junior secondary school (Grades 7 and 8) in EMI and CMI groups. They examined the impact of different MOIs (EMI versus CMI) on students' achievements. Drawing on 100 EMI and CMI schools, they recruited Grade 7 students with the same level of academic achievement prior to their admission to secondary schools and measured their science achievements from Grades 7 to 9 over a period of three years.

They compared the two MOI groups using a science achievement test, a questionnaire on their self-concept in science, and classroom observations of science lessons. The results showed that EMI junior secondary school students had lower science achievement scores and a lower self-concept than their counterpart CMI students. Yip and Tsang concluded that EMI students experience greater difficulties in understanding abstract concepts, making distinctions between scientific terms, and applying concepts in different situations through their L2. They also conducted observations in some of these EMI classrooms and concluded that EMI negatively affected the quality of science learning in these classrooms due to the students' limited English proficiency and the inadequate repertoire of instructional strategies used by their science teachers.

Moving on to senior secondary classrooms, Marsh et al. (2000) conducted a similar study with a sample of 7,802 students to examine the impact of CMI and EMI on student achievements in language and content subjects from Grades 7 to 11. The researchers concluded that in the junior secondary years, EMI has a substantial negative influence on students' self-concept and science achievement. However, the disadvantage of learning science through L2 is offset to a limited extent in senior secondary years, when students' English proficiency has improved and students have become more used to English instruction. A strong emphasis by schools on promoting English in their language lessons was also found to reduce the L2 disadvantages.

2.4.2 Research on students' academic self-concepts

'Self-concept' in science refers to students' beliefs about their academic abilities and performance in science. The formation of a positive self-concept is regarded as an important outcome of the learning process. In a language immersion context, studies of students' academic self-concept are primarily concerned with the correlation between students' self-concept and their academic achievements (Marsh et al., 2000, 2002; Yip et al., 2007a, b; Yip & Tsang, 2006; Yip, 2001). Studies have also suggested that there is a reciprocal relationship between academic achievement and self-concept (Marsh & Yeung, 1997). Pun and Jin (2022) also demonstrated that students with stronger science learning self-concepts tended to gain higher science achievement scores. They also identified a similar tendency regarding students' perceived level of English difficulty in science classrooms. In other words, once students' academic achievement has improved, their self-concept will also improve, and this leads to further improvements in their academic achievement.

Marsh et al. (2000) examined the impact of CMI and EMI on self-concepts in addition to the impact on student achievements from Grades 7 to 11. After measuring students' academic performance and self-concept in science, they found statistically that there is a strong relationship between students' lower self-concept in science and their science achievement. This also confirms the finding that students' prior achievements have an impact on their subsequent self-concept in science but that such effects do not differ between L1 and L2 instruction (Marsh et al., 2002). Although Marsh and his colleagues have provided evidence to show that EMI has a substantial negative influence on students' self-concept and science achievements at junior secondary levels, as already noted above, the disadvantage of learning science through English is offset to a limited extent at senior secondary levels, when students have gained English proficiency and have had increased exposure to full English instruction. However, as yet, there is no research that has conducted a detailed examination of senior secondary science classrooms, in terms of the kind of teaching practices and patterns of classroom interactions found in EMI classrooms. It is unclear how students' English proficiency is developed in senior secondary EMI science classrooms. There is a lack of empirical evidence about the types of teaching activities these science teachers employ during classroom interactions to support their students in acquiring science knowledge and scientific English to express the knowledge learned in EMI science classrooms.

In a longitudinal study, Yip and Tsang (2006) compared EMI and CMI groups during the first three years of junior secondary school. They found that although EMI students had better science achievements than those instructed in CMI contexts, their self-concept in science was lower than in other subjects. This suggests that EMI students experience greater learning difficulties in science, as science learning requires high levels of language proficiency to achieve abstract thinking and understanding of scientific terminology. The study also observed that EMI students perceived science as more difficult to learn than CMI students. However, they did not confirm whether or not English immersion induced an adverse effect on science students' self-concept because they failed to control for students' academic ability in the different immersion groups. Most EMI students are in Band 1 (above average academic achievements), which means they are more academically capable than students from CMI schools, who are at Band 2 (average or below-average academic achievements).

Yip et al. (2007) reported that EMI students showed higher self-concept in Chinese, English, and mathematics, but lower self-concept in science compared with CMI students. Despite their lower self-concept, the EMI students showed a greater interest in learning science than CMI students. However, while EMI students had a lower self-concept and self-competence in science, they performed better on science achievement tests than many of the CMI students. Similar to self-concept, self-competence refers to students' attitudes towards their ability to achieve academic success and fulfil academic tasks (Hayat et al., 2020).

In sum, the literature indicates that EMI students potentially face disadvantages in self-concept, language, and science achievements at the beginning of junior secondary school, but that these disadvantages are gradually mitigated when students' English proficiency has improved and they have become more used to English instruction in senior secondary school. Other studies have also shown contradictory results, including that EMI students performed better in science achievement tests despite their lower self-concept. It is therefore unclear to what extent EMI and students' limited English proficiency may negatively affect the quality of science learning in these classrooms. The next section will report research findings relevant to the language challenges EMI students experience in their classrooms and the kinds of coping strategies they adopt in addressing these challenges.

2.5   Language challenges and coping strategies in EMI classrooms

2.5.1   Challenges related to the learning process

Students in EMI science classrooms face 'double challenges' in learning science through English (Probyn, 2006; Rollnick & Rutherford, 1996). They must not only learn scientific knowledge but also English in order to read and write scientific ideas (Schleppegrell, 2004). Discourse analysts suggest that scientific language has a specific register, discourse, and lexico-grammatical features (Halliday & Martin, 1993). Students may fail to understand scientific terms and conceptualise the abstract and complex relationships between science concepts in English due to their poor English ability (Probyn, 2009; Yip & Tsang, 2006).

Research shows that even EMI students with high academic ability at the beginning of their secondary education are negatively affected in their learning during the initial years of immersion in English. English skills developed through English lessons at primary school are not sufficient for these students to master scientific content in English. At junior secondary levels, EMI students perform worse than their counterpart CMI students in science achievement tests, with problems understanding abstract concepts, distinguishing scientific terms, applying concepts in different situations, and applying learned scientific knowledge in everyday situations (Yip et al., 2003).

In the senior secondary years, the negative impact still exists but the disadvantage of learning science through English is offset to a limited extent as students have improved their English proficiency, become more familiar with EMI, or been encouraged by their school's strong emphasis on English in their English lessons. However, these science students may still be significantly disadvantaged compared with their counterparts in CMI, as well as with EMI students who study geography, history and mathematics (Marsh et al., 2000, 2002). This is because the effectiveness of EMI varies according to students' initial ability levels, meaning that students who are not particularly proficient in the English language might experience learning difficulties. Such a challenge of learning science through English could be explained by research examining the nature of school science discourse. The language demands in science are different from those in English classrooms. The language of science has a register distinct from daily language, a response to the fact that the shape and role of scientific activities are largely different from everyday human experiences (Halliday & Martin, 1993; Martin & Rose, 2008). If the unique features of scientific English are ignored, constructing and expressing scientific knowledge will be problematic (Christie & Derewianka, 2008). For instance, students may become frustrated by the ambiguity and imprecise referents of nominalisation,[1] if they fail to retrieve the hidden meanings from nominal phrases in order to gain a full understanding (Fang, 2005). An example of nominalisation is changing the verb '*neutralise*' into the noun '*neutralisation*' in the following sentences: "*An acid and a base neutralise each other. The products of neutralisation are salt and water only. Neutralisation is the reaction of an acid with a base producing salt and water only.*" Students may have difficulty in presenting science knowledge and developing logical arguments in their writing if they lack awareness of the role of nominalisation in science texts (Schleppegrell, 2004).

In sum, EMI students potentially face many language challenges in EMI classrooms. These challenges include: (1) integrating both content and language, (2) mastering the register of science discourse, (3) understanding the differences between scientific English and general English, (4) understanding abstract concepts, (5) distinguishing scientific terms, and (6) applying concepts in different situations. To overcome some of these language challenges, the following section summarises previous studies about the coping strategies that may help EMI students to tackle their language issues in their EMI classrooms.

2.5.2   *Coping strategies related to the learning process*

All programme participants demonstrated low proficiency in written English. Walker concluded that English bridging courses in Hong Kong pay limited attention to students' needs relative to their language development in content learning, or what is known as their 'cognitive academic language proficiency' (Cummins, 1979).

Walker's claim is echoed by Yip and his colleagues' study. They suggest that students need to develop both the necessary English proficiency and the capacity for L2 learning in science by filling in their current language gaps. For this to happen, two gaps must be closed. First, an effort should be made to close the gap between how students conceptualise scientific knowledge when using L2 and L1. This is achieved by the use of prompting questions to evaluate students' scientific understanding and to identify their misconceptions (Yip, 2004). Second, there is a gap between what students can *understand* in English and what they can *articulate* in English. This gap is closed by providing sufficient L2 learning opportunities through classroom interactions to test their L2 hypotheses (Lo, 2014) and by giving corrective feedback to their L2 utterances. In sum, the English enrichment courses for students switching into EMI might help prepare students with the necessary English proficiency to study content subjects in English, but a programme's effectiveness will be limited if there is a lack of attention paid to students' needs in developing their L2 disciplinary language and a failure to bridge the gaps between how students conceptualise scientific knowledge in L1 and L2.

In addition, expectations of language learning in EMI classrooms are often predicated on the belief that students' learning of English will occur through massive exposure to the L2. However, many students in English-only classrooms may not be able to gain such benefits if their English ability is too low. Therefore, it is perhaps appropriate to ask whether L1 can play an important role in EMI teaching, in light of the fact that a significant number of studies have reported the benefits of L1 in EMI teaching (Lin, 2006; Lo & Macaro, 2012; Probyn, 2006). The L1 can perhaps be beneficial in EMI contexts where content learning is the primary objective, as well as in contexts where language acquisition is also an important objective. However, very little research evidence is available on L1 use in EMI settings (Lo & Macaro, 2015) and practical pedagogical suggestions for supporting students' content and language learning (Chan, 2014; Poon et al., 2013).

2.6 Summary

As the review above has shown, the approaches and challenges pertaining to teaching and learning science through EMI are subject to debate in the Hong Kong context. At this point, it is helpful to summarise the major issues and findings that informed the research questions motivating my research project:

1 EMI science lessons seem to be teacher-centred and have limited teacher–student interactions. It is questionable whether these teacher-centred teaching practices in EMI science classrooms are the result of poor quality classroom interactions or a reflection of science teachers' inability to implement the curriculum principles. Other learning opportunities (Lo & Macaro, 2012), such as student–student interactions that are crucial to science learning have yet to be investigated.

2 EMI seems to have a negative effect on the quality of science learning because of students' limited English proficiency. However, previous studies, such as Yip

(2003) falsely assumed that students in both EMI and CMI groups acquire the same level of language ability in both Chinese and English. Recognition of the difference between L1 (Chinese) and L2 (English) and their roles in teaching and learning science opens up the research space so that we may better understand the challenges and strategies of different English ability groups with different EMI experiences.

3 EMI has a substantial negative influence on students' self-concept and science achievement at junior secondary levels (Marsh et al., 2000). However, it is not clear when and how such a disadvantage is offset at senior secondary levels and how these students' English proficiency develops.

4 There is a correlation between the inadequate English ability of students and the effectiveness of teaching, which accounts for the poor quality of science learning in EMI classrooms (Yip et al., 2000). Perceptions and attitudes of science teachers towards EMI have a significant effect on their teaching practices and classroom interactions in science. Research also indicates that the quality of teachers' explanations, their interactions with students, their questioning skills, and their language awareness of students' English needs in science seem to affect the quality of students' learning processes (Marsh et al., 2000; Yip, 2003).

5 Most research has only investigated the effectiveness of EMI on students' science achievements at junior secondary levels by comparing it with those instructed in another language medium. The research to date fails to arrive at agreed conclusions on identifying the different types of language challenges and the corresponding strategies that could or should be adopted.

This chapter has demonstrated the consequences of the adoption of EMI for teaching and learning in science classrooms using research that explores classroom interactions, teachers' and students' views on their EMI experiences, their language challenges, and their coping strategies. Hong Kong's current MOI situation, in which both CMI and EMI schools are adopting EMI, has produced a new research context yet to be explored: 'early–full' versus 'late–partial' EMI schools. Students in EMI science classrooms, particularly those from late–partial EMI schools who switched from CMI to EMI at senior secondary levels, have encountered numerous language challenges. SLA theories (Long's interaction hypothesis) and research in science education (Mortimer and Scott's framework) have argued that classroom interaction plays an important role in students' content and language learning. Thus, one of the possible ways to understand students' language challenges in these EMI classrooms would be through an investigation of the classroom interaction in both types of EMI schools, with an examination of the frequency and functions of teachers' and students' interactional contributions, teachers' questioning skills, and the choices of instructional language. These considerations informed my study's aims, research questions, and research design, which facilitated the exploration of classroom interactions and teachers' and students' perceptions of EMI in early–full and late–partial EMI secondary science classrooms.

This study explores classroom interactions in different EMI science classrooms. It compares the interaction patterns of early–full EMI[2] and late–partial EMI[3] secondary schools in Hong Kong and compares Grades 10 and 11 within each group. This project also analyses the perceptions of teachers and students with respect to their teaching and learning in the two types of schools. In order to address these issues, it is important to look at the nature and quality of classroom interactions in these EMI classrooms. Especially, the research questions addressed in this book are:

1 What is the nature of classroom interaction in EMI science classrooms?
2 What are the differences and similarities in classroom interaction between early–full and late–partial EMI schools?
3 What are the differences and similarities in classroom interaction between Grades 10 and 11 in each group?
4 What are teachers' perceptions of teaching science through EMI?
5 What are students' perceptions of learning science through EMI?

The next chapter will present the study's objectives, research questions, and research design based on the literature discussed in Chapter 2 to explore classroom interactions and teachers' and students' perceptions of EMI in early–full and late–partial EMI secondary science classrooms.

Notes

1 Nominalisation is the most common strategy used in science to theorise and construe science processes into abstract entities. This way, scientists can 'pick up' salient items from a complex sequence and 'package' them into a single semantic configuration in the subject position.
2 According to literature on Canadian immersion programmes, early immersion refers to primary schools switching to foreign language (English) instruction. With late immersion, the switch to foreign instruction starts at junior secondary schools. Very late immersion starts at senior secondary. This is why I use 'late EMI' and 'very late EMI' to refer to the EMI immersion programmes in Hong Kong. The terms 'late and very late' are problematic to readers in other countries, where they have a different meaning for the age of admission in public schools. Therefore, I use 'early–full EMI' and 'late–partial EMI' to refer to the EMI immersion programmes in Hong Kong. The term 'early–full EMI secondary schools' refers to schools that opt for English-only instruction for all content subjects from Grades 7 to 12.
3 Late–partial EMI secondary schools refer to schools that opt for Cantonese from Grades 7 to 9 and later switch to English medium from Grades 10 to 12.

References

Airey, J. (2012). "I don't teach language": The linguistic attitudes of physics lecturers in Sweden. *AILA Review*, *25*(1), 64–79. https://doi.org/10.1075/aila.25.05air

An, J., & Thomas, N. (2021). Students' beliefs about the role of interaction for science learning and language learning in EMI science classes: Evidence from high schools in China. *Linguistics and Education*, *65*, e100972. https://www.sciencedirect.com/science/article/abs/pii/S0898589821000723?via%3Dihub

Arkoudis, S. (2005). Fusing pedagogic horizons: Language and content teaching in the mainstream. *Linguistics and Education, 16*(2), 173–187. http://doi.org/10.1016/j.linged.2006.01.006

Barwell, R. (2005). Integrating language and content: Issues from the mathematics classroom. *Linguistics and Education, 16*(2), 205–218. http://doi.org/10.1016/j.linged.2006.01.002

Bernat, E., & Gvozdenko, I. (2005). Beliefs about language learning: Current knowledge, pedagogical implications, and new research directions. *TESL-EJ, 9*, 1–21. https://tesl-ej.org/ej33/a1.pdf

Chan, J. Y. H. (2013). A week in the life of a 'finely tuned' secondary school in Hong Kong. *Journal of Multilingual and Multicultural Development, 34*(5), 411–430. http://doi.org/1 0.1080/01434632.2013.770518

Chan, J. Y. H. (2014). Fine-tuning language policy in Hong Kong education: stakeholders' perceptions, practices and challenges. *Language and Education, 28*(5), 459–476. http://doi.org/10.1080/09500782.2014.904872

Chin, C. (2007). Teacher questioning in science classrooms: Approaches that stimulate productive thinking. *Journal of Research in Science Teaching: The Official Journal of the National Association for Research in Science Teaching, 44*(6), 815–843.

Christie, F., & Derewianka, B. (2008). *School discourse: Learning to write across the years of schooling*. London: Continuum International Publishing Group.

Costa, F (2012) Focus on form in ICLHE lectures in Italy: Evidence from English-medium science lectures by native speakers of Italian. *AILA Review* 25: pp. 30–47.

Coyle, D., Hood, P., & Marsh, D. (2010). *CLIL: Content and language integrated learning*. Cambridge: Cambridge University Press.

Cummins, J. (1979). Linguistic interdependence and the educational development of bilingual children. *Review of Educational Research*. http://doi.org/10.3102/00346543049002222

Curriculum Development Council and The Hong Kong Examinations and Assessment Authority (CDC & HKEAA). (2007). *Science Education Key Learning Area*. Hong Kong: CDC & HKEAA.

Dearden, J. (2015). *English as a medium of instruction–a growing global phenomenon*. Retrieved from https://www.britishcouncil.org/sites/default/files/e484_emi_-_cover_option_3_final_web.pdf

Evans, S., & Morrison, B. (2011). Meeting the challenges of English-medium higher education: The first-year experience in Hong Kong. *English for Specific Purposes, 30* (3), 198–208. https://doi.org/10.1016/j.esp.2011.01.001

Evnitskaya, N., & Morton, T. (2011). Knowledge construction, meaning-making and interaction in CLIL science classroom communities of practice. *Language and Education, 25*(2), 109–127. http://doi.org/10.1080/09500782.2010.547199

Fang, Z. (2005). Scientific literacy: A systemic functional linguistics perspective. *Science Education, 89*(2), 335–347.

Ferreira, J. G. (2011). Teaching life sciences to English second language learners: What do teachers do? *South African Journal of Education, 31*, 102–113.

Halliday, M. A. K., & Martin, J. R. (1993). *Writing science literacy and discursive power*. London: Falmer Press.

Halliday, M. A. K., & Matthiessen, C. M. I. M. (2004). *An introduction to functional grammar* (3rd ed.). London: Routledge.

Hayat, A. A., Shateri, K, Amini, M., Shokrpour, N. (2020). Relationships between academic self-efficacy, learning-related emotions, and metacognitive learning strategies with academic performance in medical students: A structural equation model. *BMC Medical Education, 20*(76). https://doi.org/10.1186/s12909-020-01995-9

Hong, J. (2022). A study of language-related episodes in online English-medium instruction classes in high schools in South Korea. *English for Specific Purposes, 67*, 65–75.

Hoare, P. 2003. *Effective Teaching of Science Through English in Hong Kong Secondary Schools.* Doctoral thesis, Hong Kong: University of Hong Kong.

Huang, J., & Normandia, B. (2008). Comprehending and solving word problems in mathematics: Beyond key words. In Z. Fang & M. J. Schleppegrell (Eds.), *Reading in secondary content areas: A language-based pedagogy* (pp. 66–83). Ann Arbor: University of Michigan Press.

Jiang, L., Zhang, L. J., & May, S. (2016). Implementing English-medium instruction (EMI) in China: teachers' practices and perceptions, and students' learning motivation and needs. *International Journal of Bilingual Education and Bilingualism, 0*(0), 1–13. http://doi.org/10.1080/13670050.2016.1231166

Krashen, S. D. (1982). *Principles and practice in second language acquisition.* Oxford: Pergamon.

Lin, A. M., (2006). Beyond linguistic purism in language-in-education policy and practice: exploring bilingual pedagogies in a Hong Kong science classroom. *Language and Education, 20*(4), 287–305. http://doi.org/10.2167/le643.0

Lin, A. M., & Wu, Y. (2015). 'May I speak Cantonese?'–Co-constructing a scientific proof in an EFL junior secondary science classroom. *International Journal of Bilingual Education and Bilingualism, 18*(3), 289–305.

Llinares, A., Morton, T., & Whittaker, R. (2012). *The roles of language in CLIL.* Cambridge: Cambridge University Press.

Llurda, E. (2013). English-medium instruction at universities: Global challenges. *ELT Journal, 67*(4), 497–500. https://doi.org/10.1093/elt/cct045

Lo, Y. Y. (2014). L2 learning opportunities in different academic subjects in content-based instruction– evidence in favour of 'conventional wisdom.' *Language and Education, 28*(2), 141–160. http://doi.org/10.1080/09500782.2013.786086

Lo, Y. Y., & Macaro, E. (2012). The medium of instruction and classroom interaction: Evidence from Hong Kong secondary schools. *International Journal of Bilingual Education and Bilingualism, 15*(1), 29–52. http://doi.org/10.1080/13670050.2011.588307

Lo, Y. Y., & Lo, E. S. C. (2014). A meta-analysis of the effectiveness of English-medium education in Hong Kong. *Review of Educational Research, 84*(1), 47–73.

Lo, Y. Y., & Macaro, E. (2015). Getting used to content and language integrated learning: what can classroom interaction reveal?. *The Language Learning Journal, 43*(3), 239–255.

Long, M. H. (1983). Native speaker/non-native speaker conversation and the negotiation of comprehensible input. *Applied Linguistics, 4*(2), 126–141.

Lorenzo, F. (2013). Genre-based curricula: Multilingual academic literacy in content and language integrated learning. *International Journal of Bilingual Education and Bilingualism, 16*(3), 375–388. http://doi.org/10.1080/13670050.2013.777391

Mak, S. Y., Yip, D. Y., & Chung, C. M. (1999). Alternative conceptions in biology-related topics of integrated science teachers and implications for teacher education. *Journal of Science Education and Technology, 8*(2), 161–169. Retrieved from http://www.jstor.org/stable/40188526

Marsh, H. W., & Yeung, A. S. (1997). Causal effects of academic self-concept on academic achievement: Structural equation models of longitudinal data. *Journal of Educational Psychology, 89*(1), 41–54. http://doi.org/10.1037/0022-0663.89.1.41

Marsh, H. W., Hau, K. T., Kong, C. K. (2000). Late immersion and language of instruction in HK high schools-achievement Growth in language and Nonlanguage subjects. *Harvard Educational Review, 70*(3), 302–346.

Marsh, H. W., Hau, K.-T., & Kong, C.-K. (2002). Multilevel causal ordering of academic self-concept and achievement: Influence of language of instruction (English compared with Chinese) for Hong Kong students. *American Educational Research Journal, 39*(3), 727–763. http://doi.org/10.3102/00028312039003727

Martin, J. R., & Rose, D. (2008). *Genre relations: Mapping culture.* Sheffield: Equinox Publishing Ltd.

Martin, J. R., & Veel, R. (1998). *Reading science critical and functional perspectives on discourses of science.* London: Routledge.

Mazak, C. (2016). 1. Introduction: Theorizing translanguaging practices in higher education. In C. Mazak & K. Carroll (Ed.), *Translanguaging in higher education: Beyond monolingual ideologies* (pp. 1–10). Bristol, Blue Ridge Summit: Multilingual Matters. https://doi.org/10.21832/9781783096657-003

Mendoza, A., & Ou, J. (2022). CACTI: Use of a survey instrument as a semistructured interview protocol to facilitate teacher retrospection on bi/multilingual practices in EMI. *System, 109*, 102887. https://doi.org/10.1016/j.system.2022.102887

Mohd Saat, R., & Othman, J. (2010). Meeting linguistic challenges in the science classroom: Pre-service ESL teachers' strategies. *Asia Pacific Journal of Education, 30*(2), 185–197. http://doi.org/10.1080/02188791003721937

Mortimer, E., & Scott, P. (2003). *Meaning making in secondary science classrooms.* Berkshire, England: Open University Press.

Othman, J., & Mohd Saat, R. (2009). Challenges of using English as a medium of instruction: Pre-service science teachers' perspective. *The Asia-Pacific Education Researcher, 18*(2), 307–316.

Poon, A. Y. K. (2013). Will the new fine-tuning medium-of-instruction policy alleviate the threats of dominance of English-medium instruction in Hong Kong? *Current Issues in Language Planning, 14*(1), 34–51. http://doi.org/10.1080/14664208.2013.791223

Poon, A. Y. K., Lau, C. M., & Chu, D. H. (2013). Impact of the fine-tuning medium-of-instruction policy on learning: Some preliminary findings. *Literacy Information and Computer Education Journal, 4*(1), 946–954.

Probyn, M. (2001). Teachers voices: Teachers reflections on learning and teaching through the medium of English as an additional language in South Africa. *International Journal of Bilingual Education and Bilingualism, 4*(4), 249–266. http://doi.org/10.1080/13670050108667731

Probyn, M. (2006). Language and learning science in South Africa. *Language and Education, 20*(5), 391–414. http://doi.org/10.2167/le554.0

Probyn, M. (2009). 'Smuggling the vernacular into the classroom': Conflicts and tensions in classroom codeswitching in township/rural schools in South Africa. *International Journal of Bilingual Education and Bilingualism, 12*(2), 123–136.

Prophet, R. B., & Badede, N. B. (2007). Language and student performance in junior secondary science examinations: The case of second language learners in Botswana. *International Journal of Science and Mathematics Education, 7*(2), 235–251. http://doi.org/10.1007/s10763-006-9058-3

Pun, J., & Jin, X. (2022). English medium of instruction in science learning: A path analysis. *System, 109*, 102867. https://doi.org/10.1016/j.system.2022.102867

Pun, J., & Tai, K. W. H. (2021). Doing science through translanguaging: A study of translanguaging practices in secondary English as a medium of instruction science laboratory sessions. *International Journal of Science Education, 43*(7), 1112–1139. https://doi.org/10.1080/09500693.2021.1902015

Rollnick, M., & Rutherford, M. (1996). The use of mother tongue and English in the learning and expression of science concepts: A classroom-based study. *International Journal of Science Education, 18*(1), 91–103. http://doi.org/10.1080/0950069960180108

Sato, M. (2013). Beliefs about peer interaction and peer corrective feedback: Efficacy of classroom intervention. *Modern Language Journal, 97*(3), 611–633. https://doi.org/10.1111/j.1540-4781.2013.12035.x

Schleppegrell, M. J. (2004). *The language of schooling: a functional linguistics perspective.* Mahwah, N.J.: Lawrence Erlbaum.

Sinclair, J. and Coulthard, M. (1975). *Towards an analysis of discourse.* Oxford: Oxford University Press.

Soruç, A., Altay, M., Curle, S., & Yuksel, D. (2021). Students' academic language-related challenges in English Medium Instruction: The role of English proficiency and language gain. *System, 103*, 102651. https://doi.org/10.1016/j.system.2021.102651

Swain M. (2009). Languaging, Agency and Collaboration in Advanced Second Language Proficiency. In Byrnes H. (Ed.), *Advanced Language Learning: The Contribution of Halliday and Vygotsky*, 95–108. London-New York: Continuum.

Tan, M. (2011). Mathematics and science teachers' beliefs and practices regarding the teaching of language in content learning. *Language Teaching Research.* http://doi.org/10.1177/1362168811401153

Tatzl, D. (2011). English-medium masters' programmes at an Austrian university of applied sciences: Attitudes, experiences and challenges. *Journal of English for Academic Purposes, 10*(4), 252–270. https://doi.org/10.1016/j.jeap.2011.08.003

Tung, P., Lam, R., & Tsang, W. K. (1997). English as a medium of instruction in post-1997 Hong Kong: What students, teachers, and parents think. *Journal of Pragmatics, 28*, 441–459.

Walker, E. (2010). Evaluation of a support intervention for senior secondary school English immersion. *System: An International Journal of Educational Technology and Applied Linguistics, 38*(1), 50–62. http://doi.org/10.1016/j.system.2009.12.005

Wannagat, U. (2007). Learning through L2 – Content and Language Integrated Learning (CLIL) and English as Medium of Instruction (EMI). *International Journal of Bilingual Education and Bilingualism, 10*(5), 663–682. https://doi.org/10.2167/beb465.0

Wellington, J., & Osborne, J. (2001). *Language and literacy in science education.* Philadelphia, USA: Open University Press.

Wenger, E. (1998). Communities of practice: Learning, meaning, and identity. *Systems thinker* (Vol. 9). Cambridge: Cambridge University Press. http://doi.org/10.2277/0521663636

Yip, D. Y. (1999). Implications of students' questions for science teaching. *School Science Review, 81*(294), 49–53.

Yip, D. Y. (2001). Promoting the development of a conceptual change model of science instruction in prospective secondary biology teachers. *International Journal of Science Education, 23*(7), 755–770. http://doi.org/10.1080/09500690010016067

Yip, D. Y. (2003). *The effects of the medium of instruction on science learning of Hong Kong secondary students.* Doctoral dissertation: University of Nottingham.

Yip, D. Y. (2004). Questioning skills for conceptual change in science instruction. *Journal of Biological Education, 38*(2), 76–83. http://doi.org/10.1080/00219266.2004.9655905

Yip, D. Y., Coyle, D., & Tsang, W. K. (2007a). Evaluation of the effects of the medium of instruction on science learning of Hong Kong secondary students: Instructional activities in science lessons. *Education Journal, 35*(2), 77–107.

Yip, D. Y., Coyle, D., & Tsang, W. K. (2007b). Medium of instruction on science students: Instructional activities in science lessons. *Education Journal, 35*(2), 77–107.

Yip, D. Y., & Tsang, W. K. (2006). Evaluation of the effects of the medium of instruction on science learning of Hong Kong secondary students: Students' self-concept in science. *International Journal of Science and Mathematics Education, 5*(3), 393–413. http://doi.org/10.1007/s10763-006-9043-x

Yip, D. Y., Tsang, W. K., & Cheung, S. P. (2003). Evaluation of the effects of medium of instruction on the science learning of Hong Kong secondary students: Performance on the science achievement test. *Bilingual Research Journal, 27*(2), 295–331. http://doi.org/10.1080/15235882.2003.10162808

Yu, R. D. (2002). Integrating language and content: How three biology teachers work with non-English-speaking students. *International Journal of Bilingual Education and Bilingualism, 5*(1), 40–57. http://doi.org/10.1080/13670050208667745

3   Research methods

3.1   Research design in detail

The study adopts a mixed-methods approach, employing a cross-sectional (2×2) design to investigate whether there were any significant differences in classroom interactions, teacher and student perceptions between full and partial EMI secondary schools, and differences and similarities between Grades 10 and 11 in each group. The research design, including the relationship between different sampling groups and the research questions, is presented in Figures 3.1 and 3.2. The research reported in this book is of a recent large-scale study that explored the teaching and learning processes of EMI senior secondary science classrooms (physics, chemistry, and biology) in 30 secondary schools in Hong Kong.

This study has three parts. The first is a case study of the nature of classroom interaction in EMI science classrooms. The case study compares classroom interaction patterns between early–full and late–partial EMI schools using video-recorded observation data of classes. This component of the methodology addresses the first three research questions (reiterated in Figure 3.1) by identifying similarities and differences in the classroom interactions of the two types of schools, both in general and specifically at Grades 10 and 11.

The second component is an investigation of the participants' views on EMI instruction using interview data (i.e., interviews with the teachers and students). This component addresses research questions 4 and 5 (reiterated in Figure 3.2), by identifying recurrent issues from the teachers' and students' interviews about their views, attitudes, and personal experiences of teaching and learning science through English in both types of schools.

The third component also addresses research questions 4 and 5 by exploring the relationships among the variables related to teachers' and students' perceptions using survey data (i.e., teacher and student questionnaires). The variables include participants' choice of classroom language, their thoughts on teaching and learning science through the medium of English, the language challenges of doing so, coping strategies in EMI science classrooms, and other learning outcomes such as the students' self-concept in science and self-reported academic achievements in English

DOI: 10.4324/9781003001454-4

Observations

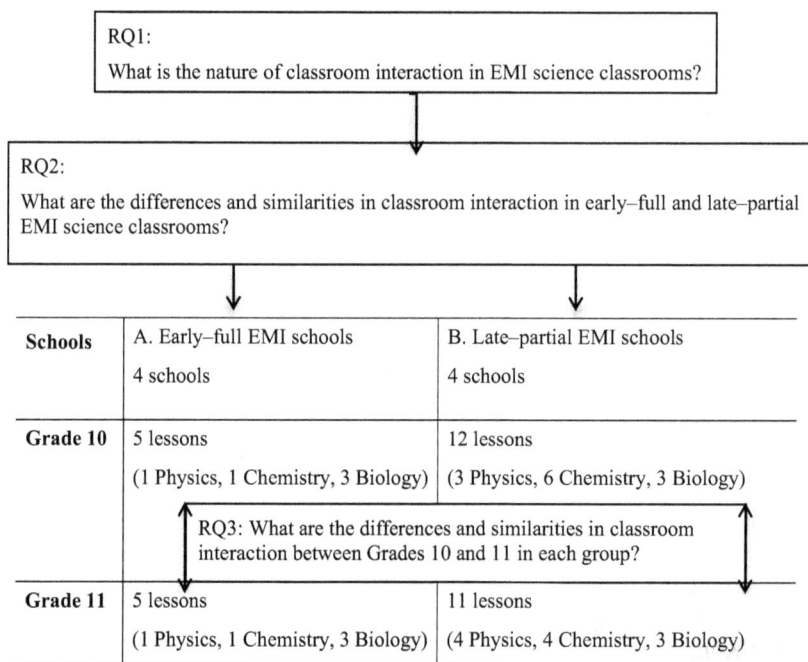

RQ1:
What is the nature of classroom interaction in EMI science classrooms?

RQ2:
What are the differences and similarities in classroom interaction in early–full and late–partial EMI science classrooms?

	A. Early–full EMI schools	B. Late–partial EMI schools
Schools	4 schools	4 schools
Grade 10	5 lessons	12 lessons
	(1 Physics, 1 Chemistry, 3 Biology)	(3 Physics, 6 Chemistry, 3 Biology)
	RQ3: What are the differences and similarities in classroom interaction between Grades 10 and 11 in each group?	
Grade 11	5 lessons	11 lessons
	(1 Physics, 1 Chemistry, 3 Biology)	(4 Physics, 4 Chemistry, 3 Biology)

Figure 3.1 Research design (classroom interactions).

Interviews, Questionnaires

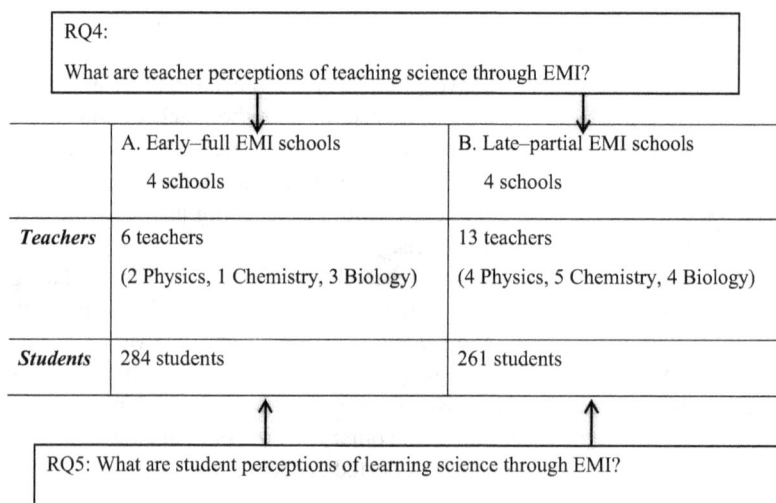

RQ4:
What are teacher perceptions of teaching science through EMI?

	A. Early–full EMI schools	B. Late–partial EMI schools
	4 schools	4 schools
Teachers	6 teachers	13 teachers
	(2 Physics, 1 Chemistry, 3 Biology)	(4 Physics, 5 Chemistry, 4 Biology)
Students	284 students	261 students

RQ5: What are student perceptions of learning science through EMI?

Figure 3.2 Research design (teachers' and students' perceptions).

Table 3.1 Summary of the research aims, methods and analysis[1]

Research question	Research instruments	Source of data	Method of analysis
1. What is the nature of classroom interaction in EMI science classrooms? 2. What are the differences and similarities in classroom interaction between early–full and late–partial EMI schools? 3. What are the differences and similarities in classroom interaction between Grades 10 and 11 in each group?	Classroom observations	Transcripts of the video-recorded lessons, the observation notes, Grade 10 and 11 EMI science classrooms ($n = 33$ observations)[1]	Discourse analysis of classroom interactions using EMI science classrooms
4. What are teacher perceptions of learning science through EMI? 5. What are student perceptions of learning science through EMI?	Questionnaire Interviews	Survey data ($n = 19$ teachers; $n = 545$ students) Interview data (pre- and post-observation) ($n = 19$ teachers; $n = 52$ students)	Mean-comparison scores of teacher perceptions and student perceptions of teaching and learning in different groups Factor analysis to identify the major factor loading among the survey items related to language challenges and coping strategies. Major factors are further examined by t-test/ Mann–Whitney test to determine significant group differences.

and science. My exploration of the possible relationships among these variables identifies differences in teacher and student attitudes towards EMI instruction in the two types of schools. It also suggests the extent to which the students' academic achievements in English and science are related. Table 3.1 summarises the study's aims, research instruments, data sources, and analytical methods.

3.2 Justification for a mixed-methods approach

Data was collected from the classrooms and analysed using a range of methods. These included qualitative data (i.e., semi-structured interviews and 33 video-recorded classroom observations) to explore the similarities and differences in classroom interactions during the first and second years of the senior science curriculum (Grades 10 and 11) in the two types of EMI schools. The research design adopted a multiphase mixed-methods approach, interactively combining qualitative and quantitative components across all stages of the research design. This methodology provides a continuous 'dialogue' between the quantitative and qualitative approaches in various stages of the research process, including instruments, data collection, and integration of analyses and findings (Kington et al., 2011).

The combination of quantitative and qualitative approaches is more likely to offer a comprehensive understanding of the phenomena under investigation. This is because mixed methods can overcome the weaknesses of either a qualitative or quantitative approach used on its own. In particular, the quantitative approach has been criticised as simplistic, de-contextualised, and unable to capture the true meaning of the data, whereas the qualitative approach has been criticised as being very context-specific, using samples that are not representative enough (Dörnyei, 2007). Combining both approaches can therefore compensate for the weaknesses of each and increase the validity of the results (Dörnyei, 2007).

The study follows an exploratory-sequential design (Creswell & Plano Clark, 2011), in which the qualitative components were followed by quantitative components. The method involved five stages, which are detailed below.

In the first stage, qualitative and quantitative instruments were developed based on an intensive literature review of empirical studies relevant to this study. In the second stage, qualitative data (observations and semi-structured interviews) were collected and analysed. Classroom interactions were observed in both early–full and late–partial EMI schools, and in both Grade 10 and 11 classrooms. The interactions were coded and analysed quantitatively according to (1) the type of interactive sequences (i.e., teacher–student, student–student) and sequence patterns (IRF or IRFRF), (2) the choice(s) and duration of the instructional language, and (3) the frequency and type of teacher and student questions.

In the third stage, emerging findings from the qualitative data (e.g., background information about the context) were used to refine the items covered in a larger-scale quantitative questionnaire. These qualitative findings from observations and interviews also informed the content validity of the quantitative survey. The survey provided mean-comparison scores of teacher and student perceptions of EMI science learning and a list of factor loadings of different language challenges and coping strategies during interactions.

In the fourth stage, analysis of the questionnaire results provided more guided questions for follow-up interviews with teachers and students. Any noticeably consistent or inconsistent views between respondents' perceptions and their actual classroom behaviour were identified.

Finally, data collected at different stages underwent a triangulation process to ensure the validity of the study. This involved checking the consistency of findings generated by the different data collection methods. As Patton (1999, p. 1192) points out,

> Studies that use only one method are more vulnerable to errors linked to that particular method (e.g., loaded interview questions, a biased or untrue response) than are studies that use multiple methods in which different types of data provide cross-data validity checks.

This requires comparing and cross-checking the consistency of information collected at different stages and by different means within both the qualitative method and the quantitative method, for instance, comparing observational data with interview data, and comparing the quantifiable data from qualitative classroom interaction patterns with the survey data about teaching and student perceptions of EMI learning, language challenges, and coping strategies.

Patton (1999) notes that triangulation of data sources within a single method seldom leads to a single and consistent result because different kinds of data have captured different results. Such discrepancies do not mean that the data is invalid but offer an opportunity for researchers to explore the reasons for the differences. However, consistency in the overall patterns of data from different sources, such as observational data and interview data in the qualitative method, as well as reasonable explanations for any discrepancy between them, are essential to the overall credibility of the findings.

Based on the quantitative survey and qualitative data from observations and interviews, group differences in classroom interactions and respondent perceptions were identified. A small sample of these classrooms with notably inconsistent views between perception and actual interactions were selected for an in-depth case study, which used discourse analysis to examine the interaction patterns. Observational notes and interview data supplemented the discourse analysis and provided information on background and context (i.e., learning targets, the researcher's comments on the observed lesson, the teachers' justifications for a bilingual pedagogical act or use of a code-switching strategy in a teaching episode). The case study of classrooms with inconsistencies between perception and interaction generated additional research questions or hypotheses that were further examined in the subsequent analyses using the larger sample of quantitative data.

Figure 3.3 shows an overview of the adopted mixed-methods approach. The figure illustrates how the qualitative components are linked to the quantitative components at different stages of the study. The diagram also highlights the five stages in the research process.

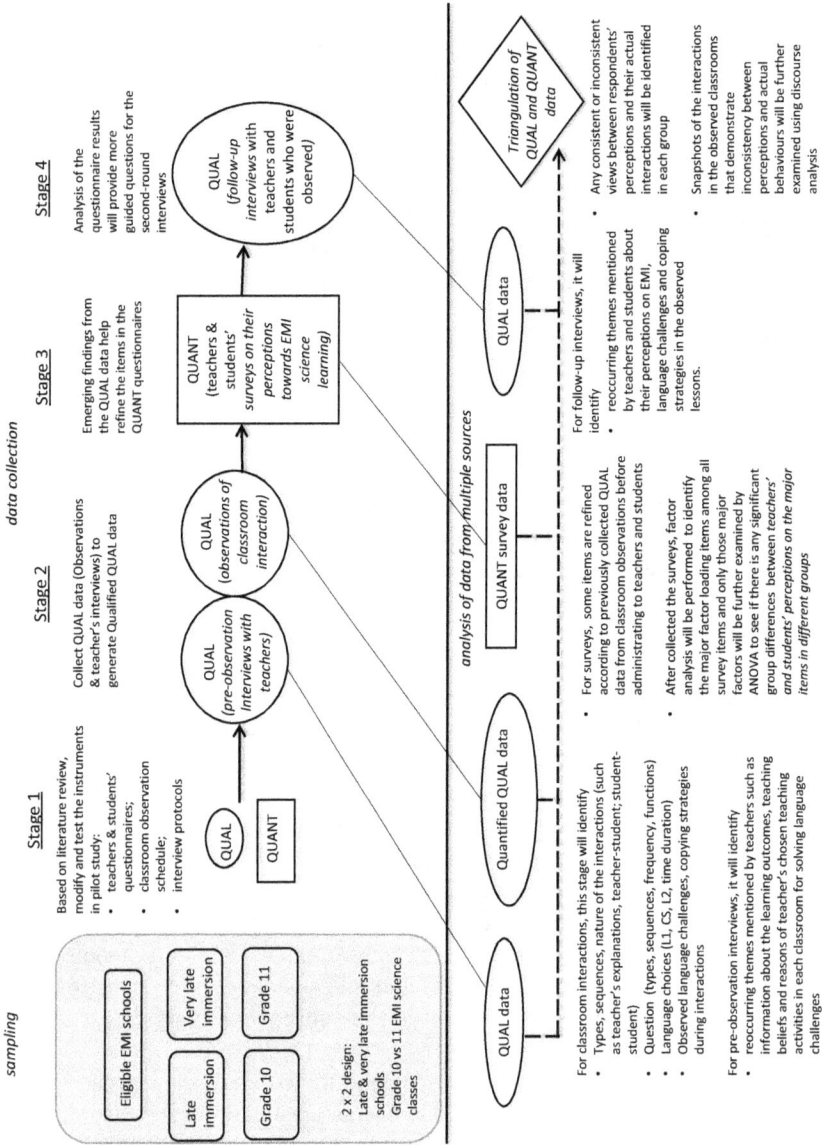

Figure 3.3 The five stages in this research process.

3.3   Data collection

3.3.1   *The population*

The 485[2] secondary schools in Hong Kong that deliver six years of secondary education according to the Hong Kong Diploma of Secondary Education (HK-DSE) curriculum form the population used here. This study examines two types of these schools: (1) early–full EMI schools (EMI instruction from Grades 7 to 12 for most content subjects) and (2) late–partial EMI schools (Cantonese instruction from Grades 7 to 9 and English instruction from Grades 10 to 11 for most content subjects).

The participants are teachers and students from Grade 10–11 EMI science classrooms (physics, chemistry, and biology) in early–full and late–partial EMI secondary schools in Hong Kong. Students are 15–18 years old and have completed six years of primary and three years of junior secondary compulsory education. Regardless of their MOI at junior levels, students in both groups have studied integrated science in Grades 7–9. During the data collection stage, students were studying the HKDSE science curriculum in Grades 10–12. These students had already received three to four years of integrated science training, covering topics, including scientific investigation, life and living, the material world, energy and change, Earth and the relationship among science, technology and society (CDC, 2002).

This study examines teacher and student perceptions of their EMI experience, particularly their classroom interactions, during the first and second years of the senior curriculum. This senior secondary period (Grades 10 and 11) is considered especially significant, as these students are at the early stage of two major transitions. First, they have moved from the junior science curriculum to the senior curriculum. Previously, these students had studied science in one integrated subject (i.e., 'science'), but in Grades 10 and 11, they are streamed into different specialised science disciplines (i.e., physics, biology, chemistry). Second, science students in Grades 10 and 11 at late–partial EMI schools have to switch from Cantonese to English as the medium of instruction.

Teachers selected from those teaching the senior science curriculum in the two types of schools had a bachelor of science degree in their teaching area and a postgraduate diploma in secondary education. All the selected teachers delivered science content that matched their disciplinary training in science, as the study design tried to ensure that the participating teachers had adequate pedagogical content knowledge and were familiar with the teaching curriculum in the HKDSE senior science curriculum (see further CDC & HKEAA, 2007a, 2007b, 2007c). Table 3.2 summarises the information on the schools' MOI and their years of exposure to EMI.

Table 3.2 Summary of the schools' MOI and years of experience in EMI

Schools	Early–full EMI schools		Late–partial EMI schools	
When did EMI start?	Grades 7–12		Grades 10–12	
Other medium of instruction	Only English, except for Chinese language subject, Chinese history		Cantonese: Grades 7–9	
MOI of science subjects	English (Grade 7–9 integrated science) English (Grade 10–12 physics, chemistry, biology)		Cantonese (Grade 7–9 integrated science) Cantonese and English (Grade 10–12 physics, chemistry, biology)	
Grades	Grade 10	Grade 11	Grade 10	Grade 11
Years of experience in EMI instruction	4	5	0	1

3.3.2   *The sampling frame*

A sampling frame was developed to ensure the balance between the generalisability and validity of the chosen sampling. The sampling frame considered:

1 The representativeness of the selected school in reflecting a typical secondary school in Hong Kong. Schools were chosen only from the districts where students had similar and homogeneous demographic backgrounds (e.g. New Territories).
2 Inspection of the student cohort's academic achievement[3] during the year of admission to the school. Only schools that had admitted students with above-average academic achievements were included.
3 Other confounding variables of individual students, such as previous educational background, family education and socioeconomic status (SES), students with behavioural difficulties, international schools or those with curricula other than the HKDSE (i.e., International Baccalaureate Diploma Programme schools) were not included.

3.3.3   *The sample*

Based on the above criteria, I investigated which schools would be eligible for this study. First, I examined the government's school profile list, 'Secondary School Profile 2013–14' (Committee on Home-School Co-operation, 2014). Second, I visited school websites for further details about their student population at each grade, the SES of students, the science subject(s) offered, the nature of the science curriculum at senior levels, and the number of science teachers.

With this information, I identified several eligible schools from early–full and late–partial EMI schools in different areas of Hong Kong. I then referred to the government's annual statistical report, which summarises demographic information for each district in Hong Kong, including median monthly household income, major occupations of residents, and educational attainment level of the residents (Census and Statistics Department, 2014). Eight districts have residents who share similar demographic characteristics. These districts are located in different areas of Hong Kong: the New Territories, Kowloon, and Hong Kong Island.

In Hong Kong, 450 of the total 485 secondary schools (92.8%) deliver the HK-DSE science curriculum. After eliminating those that did not offer the curriculum and those that admitted students with average or low academic achievement (Band 2 and 3 schools), 173 schools remained (35.7%). Among the remaining schools with above-average students (Band 1 schools), 118 were early–full EMI schools that used English as the medium of science instruction, whereas 55 were late–partial EMI schools, which switched from Cantonese to English medium at Grade 10. From there, 6 out of 55 late–partial EMI schools were eliminated as they offered curricula with either CMI or only arts subjects. In all, 118 full EMI schools (24.3%) and 49 late–partial EMI schools (10%) in the 8 districts in Hong Kong were eligible for recruitment. Figure 3.4 summarises the school selection process.

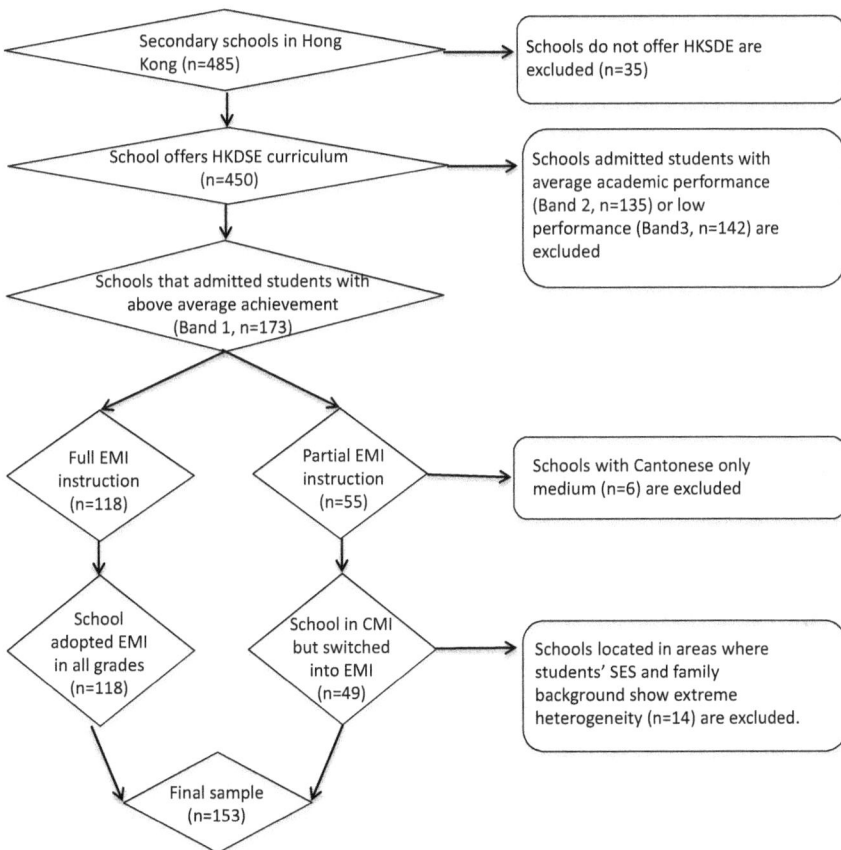

Figure 3.4 Flowchart of the school selection process.

3.3.4   Recruitment

Based on the complete list of 118 early–full EMI schools and 49 late–partial EMI schools, I contacted each school by email or telephone to invite them to participate in the study. Once I received the approval of the school principal, I contacted eligible teachers by telephone or email to introduce myself and explain the purpose of my study. With those who agreed to participate, I discussed which science topic I could observe and confirmed the schedule for observations and interviews. On the day of data collection, I gave a general briefing about the study to the teachers and their students, reminding them that this study was voluntary and informing them that those who did not want to participate could leave at any stage and their data would be destroyed.

3.3.5   Challenges of getting access to schools

During the main data collection phase, I encountered several problems getting access to secondary schools during the first two months including:

1   I did not have local connections with school principals. Most schools that agreed to participate were facilitated through personal contacts, but the rejection rate from the 153 eligible schools was high.
2   There appears to be little interest in educational research in Hong Kong secondary schools.
3   Teachers were not keen to be audio- or video-recorded. This was the primary reason for the teachers' rejection of my invitation.
4   School principals and teachers expected additional support (e.g., professional development workshops for teachers) from the researcher in return for participating in the study.

3.4   Participants

3.4.1   The eight participating schools

A letter inviting participation in the study was mailed from the researcher to the school principals and science panel heads of the 114 eligible schools. Forty schools replied, but only 20 schools expressed their interest. The response rate was 17.5%. Of these 20 schools, only 8 schools offered both Grade 10 and 11 science classes for video-recorded observations. The eight schools were chosen according to the sampling frame described above. There were four full EMI schools and four late–partial EMI schools. The success rate of getting schools to participate in this study was 7%.

The participating schools in this study included eight secondary schools in five districts in Hong Kong with similar demographic backgrounds. These sampled schools were carefully selected based on the planned sampling frame[4] (see Section 3.2.2) and according to (1) the school's banding, (2) the school's official MOI,

(3) geographical location, (4) students' academic performance, and (5) school curriculum. The selection was intended to ensure that all the sampled schools were representative of Hong Kong secondary schools and that they were closely matched in SES and students' academic ability. Figure 3.5 illustrates the locations of the 8 participating schools, as well as 20 other schools that agreed to participate but were excluded from the study.

3.4.2 School types: 'Early–full' EMI versus 'late–partial' EMI schools

With the above sampling criteria, 114 schools in 18 districts were identified in the sampling frame. Among these schools, 4 'early–full' EMI schools (identified as A, B, C, D) and 4 'late–partial' EMI schools (identified as E, F, G, H) participated in this study.

The EMI schools A to D adopted English as the only MOI for all grades (Grades 7–11) and for nearly all subjects except Chinese history and Chinese. Students in these EMI schools had passed the level of English proficiency required by the schools for studying in English. Schools E–H are late–partial EMI schools. They used Chinese as the MOI for the junior levels (Grades 7–9), and introduced partial EMI instruction at senior levels (Grades 10 and 11) to all content subjects

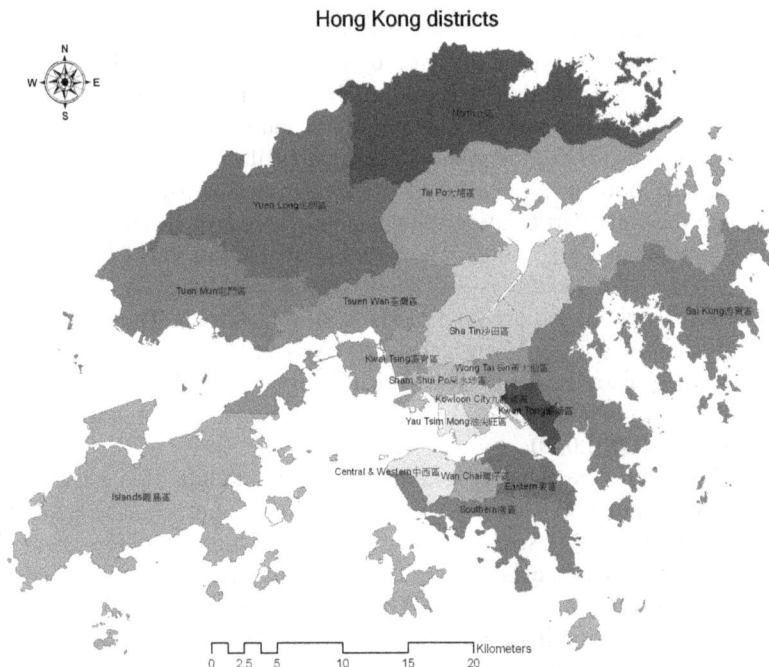

Figure 3.5 Locations of the responding schools and eight participating schools.

taught in English (including physics, chemistry, and biology). Students subsequently took a public examination in English.

Under the government's fine-tuning language policy reviewed in Chapter 2, schools E–H have been allowed to adopt EMI for senior levels if the students' English proficiency has improved to the threshold level for learning in an English environment (Legislative Council, 2009). These late–partial EMI schools used Cantonese as the MOI from Grades 7 to 9 and later switched to EMI for Grades 10–12, with nearly all content subjects being taught in English and public examinations conducted in English. The decision on the MOI was directly related to the academic ability of the students admitted to the schools.

3.4.3   School background

The background information on the participating schools is summarised in Table 3.3. This includes (1) school district, (2) school type, (3) religion, (4) median monthly household income in the school district, (5) proportion of the non-student population aged 15 or above with a post-secondary qualification, (6) academic achievement of intake students at Grade 7, and (7) the perceived language ability of intake students (i.e., the medium of instruction). This information was obtained from multiple sources, including the schools' websites, the Education Bureau's website, and interviews with the schools' principals and science teachers.

All eight participating schools were government-funded public schools. All such schools in Hong Kong are operated by the government, charities, or religious groups. Regardless of their religious affiliation (Protestant, Catholic, or no religion) or the composition of the student population (boys only, girls only, or co-ed), these schools deliver the same HKDSE curriculum and are administered by the Education Bureau of the HKSAR (equivalent to the Ministry of Education). Among the eight participating schools, there were five co-ed schools, two girls' schools, and one boys' school. They were located across various districts in Kowloon and the New Territories—Kowloon City, Kwun Tong, and Kwai Tsing had one full EMI school and one late–partial EMI school each. The two remaining schools were in Yau Tsim Mong (early–full EMI) and Tsuen Wan (late–partial EMI). Most of the schools were in areas with an average monthly income of about HKD$23,000, except for schools A, E, and H, which were located in slightly more affluent areas (HKD$31,000).

All these schools were Band 1, meaning that they had admitted a large percentage (70–100%) of students categorised as Band 1 (see note 3). The academic ability of students in EMI schools C and D was the highest (100% Band 1), closely matched by EMI schools A and B (90% Band 1) and followed by the MOI-switching schools (80% in schools F and G; 70% in schools E and H).

Among the MOI-switching schools, schools F and G were fairly comparable to the EMI schools. Students in school F were slightly more proficient in English than those in school G. Students in schools F and G were generally better at English than students in schools E and H. Such observable discrepancies

Table 3.3 The demographic backgrounds of the eight participating schools

Schools	Early–full EMI schools				Late–partial EMI schools			
	A	*B*	*C*	*D*	*E*	*F*	*G*	*H*
Actual MOI by observation	Mainly English	Mainly English	Mainly English	Mainly English	Mixed, majority English	Mixed, majority Cantonese	Mixed, majority Cantonese	Mixed
Location (district)	Kowloon City	Kwun Tong	Kwai Tsing	Yau Tsim Mong	Kowloon City	Kwun Tong	Kwai Tsing	Tusen Wan
Gender; school type (funding)	Co-ed/aided&	Co-ed/aided	Co-ed aided	Girls/aided	Boys/aided	Girls/aided	Co-ed/aided	Co-ed/aided
Religion	Protestant/Christian	Catholic	Protestant/Christian	Protestant/Christian	Catholic	Catholic	Protestant/Christian	None
Academic achievement of intake students at Grade 7[a]	90% Band 1	90% Band 1	100% Band 1	100% Band 1	70% Band 1	80% Band 1	80% Band 1	70% Band 1
Perceived MOI of intake students[b]	English	English	English	English	English and Chinese	English and Chinese	English and Chinese	English and Chinese
Median monthly household income in the district (HKD)[b]	$31,000	$23,200	$23,600	$30,000	$31,000	$23,200	$23,600	$31,000
Proportion of non-student population aged >15 with a post-secondary qualification	84.6%	75.8%	74.7%	84.1%	84.6%	75.8%	74.7%	83%

[a] 2014 Population and Household Statistics Analysed by District Council District, Census and Statistics Department, Hong Kong SAR Government and school websites, interviews with school principals and science panel head.

[b] (Median monthly domestic household income in Hong Kong for 2014: HKD$23,500) Aided: schools are subsidised by the Hong Kong SAR government

in students' standard of English may help to explain the difference in inter-action patterns observed after the schools switched from Chinese to English for teaching science. While the participating early–full and late–partial EMI schools claim English as their official MOI in Grades 10–12, it was assumed that teachers and students have gone through years of different experiences in English instruction, resulting in varied levels of academic ability and English proficiency.

3.4.4   The participating teachers

A total of 19 teachers and 545 students from Grade 10 and 11 EMI science classes were recruited from 'early–full' EMI schools (full EMI instruction from Grades 7 to 12) and 'late–partial' EMI schools (Chinese medium from Grades 7 to 9 and partial EMI instruction from Grades 10 to 12). Students in 16 of the classrooms had studied in CMI up to Grade 10, while those in 15 other classrooms had studied in English since Grade 7. Principals or science panel heads of the sampled schools invited their science colleagues to participate, provided that the teacher was teach-ing Grade 10 or 11 science subjects (physics, chemistry, and biology) and agreed to be video- or audio-recorded during classroom observation. Only eight schools were studied. These schools provided at least two science classes, both Grades 10 and 11, for classroom observations.

3.4.5   Teacher backgrounds

Nineteen teachers (3 females and 16 males) teaching science subjects (6 physics, 6 chemistry, and 7 biology) participated in this study. Each participating teacher was given a pseudonym and a code (T1–T19), which is used in the data analysis sections. Their teaching experience ranged from 3 to more than 35 years. All of the teachers graduated from universities in Hong Kong and obtained a BSc degree in a relevant teaching discipline and a postgraduate diploma in education. One teacher (T9) had a master's degree in science education. At present, there is no standardised English assessment for content-subject teachers in Hong Kong to teach at an English medium school.

3.4.6   Teacher qualifications

Based on the responses from the questionnaires and teacher interviews, the self-reported English levels of participating teachers were summarised. Almost none of the participating teachers had any formal qualifications in teaching science in English or had participated in any EMI teacher training. Only one teacher (T6) indicated in the questionnaire that she had attended a two-week professional de-velopment workshop on language across the curriculum at a local university. All the teachers had passed their English language subject before being admitted to a university programme, and they had an English proficiency equivalent to a score six in the International English Language Testing System examination. In the questionnaire, the researcher asked each teacher to rate his or her own English proficiency in listening, speaking, writing, and reading on a five-point Likert scale. However, the self-reported proficiency in English language skills might not truly reflect the actual English language abilities of participating teachers when explaining science to students and delivering teaching materials in EMI classrooms.

One issue encountered is that it was not possible to measure the teachers' English proficiency using a standardised test. Therefore, the research used direct observation to estimate the teachers' English proficiency. Little evidence was observed to suggest that teachers' English proficiency did affect their interaction with students. The inability to measure the teachers' proficiency more formally could have constituted a potential limitation of the study. However, the level of the teachers' English proficiency was not related to the study's objectives and did not address any of the research questions, namely, to examine if there was any difference in classroom interactions. In reality, teachers often found it offensive if they thought that their English proficiency was being assessed.

3.4.7 Teaching allocations

From the researcher's view, having the same teacher teaching both grades would have been the most desirable situation. This could have alleviated the potential influence of differences in teaching styles between individual teachers. It would have also been ideal if all participating science teachers had taught the same topics to minimise any possible teacher effects. However, the possibility of these circumstances would have been highly dependent on the schools' staffing, teachers' involvement, and their teaching schedules. In reality, it was not always possible to observe the same teacher teaching both Grades 10 and 11 in the same school. Only teachers in the early–full EMI school B and late–partial EMI schools G and H were teaching both Grade 10 and 11 science classrooms. Teachers in the other five participating schools were only teaching either Grade 10 or Grade 11 science.

3.4.8 The participating students

A total of 796 students aged 15–17 participated in this study—363 Grade 10 students and 428 Grade 11 students from 25 classes (6 physics, 8 chemistry, and 11 biology). Most of the classes ranged between 25 and 40 students in size.[5] The student questionnaire indicated that nearly all students were born in Hong Kong, with a small percentage of immigrants (10%) from Mainland China. All participating students indicated that they spoke Cantonese as their first language, regardless of the MOI of their schools or their place of birth.[6] Grade 10 and 11 students from the participating schools taking science subjects (physics, chemistry, or biology) were selected for observation. In each school, at least one Grade 10 and one Grade 11 class of the same science subjects were included. In schools B, G, and H, Grade 10 and 11 science classes were taught by different science teachers in each school.

3.4.9   *English ability and academic performance of students*

Participating students were asked in a questionnaire about their English proficiency, based on their school examination results from the previous academic year. Scores were confirmed with teachers to ensure the reliability of self-reported academic achievements. After I collected all students' questionnaires, I gave the overall scores of the whole class to the teachers and asked whether the scores reflected the students' academic performance at the class level. In most cases, the teachers commented that they did. I wanted to find out whether students' English achievements were related to their science achievements at school. Thus, the reported results based on their own school assessments could have sufficiently addressed the research question. However, one might argue that student English proficiency would be more accurately measured using a standardised test.

Table 3.4 shows the academic abilities of school graduates in the eight participating schools. The table contains information about student pass rates in English in public examinations, the percentage of students meeting the minimum requirement for university entry, the percentage of students enrolled in a university degree programme, the academic achievement of intake students at Grade 7, and the perceived language ability of intake students in each school.

This information showed that the students' academic achievements (e.g., science, mathematics) within each of the eight participating schools were generally similar, and this was reflected in several ways. First, the school's banding partly indicated students' English proficiency. Students are allocated to a banding based on a standardised school attainment test at Grade 6 of their admission to secondary school. Students from Band 1 EMI schools had the highest academic attainment and English language proficiency (see note 3 for further elaboration). Thus, the banding of the participating school could indicate the school's general academic ability and the English proficiency of the students admitted to it.

Secondly, the academic performance (including English proficiency) of Grade 12 graduates in public examinations provides another indicator of the student's academic ability in each school. The public examination results of graduates from the eight participating schools were above the minimum requirement for university entry in 2014. Student pass rates in the English language in public examinations (in Grade 12 in 2014) ranged from 96.9% to 100% in the EMI schools and 51% to 93.7% in the MOI-switching schools. In all the EMI schools and two of the MOI-switching schools (F and G), the pass rate was higher than the mean average of all Hong Kong secondary schools (78%).

Table 3.4 Academic ability of graduates (Grade 12) in the participating schools[a,b] 78.1% for all secondary schools in Hong Kong

	Schools	Actual MOI by observation	Student pass rate of English language in public examinations in 2014[a,b]	Proportion of students meeting the minimum requirement for university entry in 2014	Proportion of students enrolled in university degree programme in 2014[a](%)
Early–full EMI	A	English	96.9%	78.9%	45.3%
	B	English	96.9%	Not given	Not given
	C	English	100%	93.5%	71.5%
	D	English	99.3%	Not given	Not given
Late– partial EMI	E	Mixed	60%	Not given	Not given
	F	Mixed	93.7%	75.3%	Not given
	G	Mixed	75.1%	57.8%	30.7%
	H	Mixed	51%	40%	14.6%

[a] School websites and interviews with school principals and science panel heads
[b] 2014 Population and Household Statistics Analysed by District Council District, Census and Statistics Department, the Hong Kong SAR Government;

3.5 Descriptions of research instruments

3.5.1 Classroom observations

The qualitative component consists of detailed unstructured descriptive field notes of the observed lessons. The notes include information on the structure of a lesson, the teachers' pedagogical moves, the flow of interactions, the classroom climate, and other non-verbal behaviour. Observations were of single-period lessons in two visits of 30–45 minutes, or (more commonly) of double-period lessons in a single visit. The video-recorded observations were further analysed in terms of the patterns of classroom interactions. The analytical framework and coding procedures are further explained in Section 3.8.1.

3.5.2 Pre-and post-observation interviews

Pre-observation interviews were conducted with teachers at Stage 2. Observations and semi-structured interviews were analysed (see Figure 3.3), seeking information on learning outcomes, targets, teaching beliefs, and language issues related to science teaching in EMI classrooms. At the end of each observation, a post-observation interview with teachers and students was conducted at Stage 3. These qualitative findings from observations and interviews informed the content validity of the quantitative survey (see Figure 3.3). These interviews investigated participants' reflections on the classroom interactions I had just observed and asked for teachers' comments in relation to the pedagogical function of the interactions in teaching or learning science through the medium of English. Individual language challenges and coping strategies were also explored.

The interview questions were adapted from previous empirical studies, with validated instruments (Hoare, 2003; Wong, 2012) adapted to meet this study's aims. The questions were aimed at examining teacher and student views on teaching practices, language challenges, and the strategies used by the teachers when resolving any language challenges encountered by their students in understanding and communicating scientific ideas in EMI science classrooms. The interview questions were checked by three experienced researchers in the field of classroom research for content validity.

Post-observational interviews were conducted with all science teachers and with a randomly selected small sample of students (15–20 at each school) after the observation lessons. These interviews aimed to seek clarification and reasons for each teaching or learning episode where I observed language challenges and strategies in EMI science classrooms. I explored the teachers' turn-taking styles, interactions, questioning skills, code-switching patterns, and the use of multimedia for teaching science. Teachers and students were interviewed in their preferred language, Cantonese. I interviewed each teacher separately to enable them to freely express their opinions. With the students, I formed focus groups of two or three. Respondents were first asked for their comments on the observed lesson—whether they believed the lesson was typical of their previous lessons. This helped to establish the content validity of the observed data, clarifying the extent to which the observed lessons provided evidence of the actual daily experience of teaching science in those students' EMI classrooms. All interviews were audiotaped and later transcribed into Chinese for coding purposes. Only those yielding results were translated into English. The coding framework and procedures are further explained in Section 3.7.2.

3.5.3   *Teacher questionnaire*

The questionnaire design was based on a comprehensive literature review of relevant empirical studies, and a careful examination of validated surveys in previous studies of teacher and student perceptions of EMI learning. The modified survey in this study focused on aspects of significant indicators that might influence the quality of classroom interactions (Lo & Lo, 2014), including classroom language use (Evans, 2000), teachers' views and teaching practices about second language teaching in science (Hoare, 2003), and language challenges and strategies in EMI classrooms (Tatzl, 2011). The following items were covered in the teacher questionnaires:

1 Demographic data about the science teacher, teaching qualifications, teaching experiences in EMI, views on English language skills in science, and self-reported English language abilities in four language skills (speaking, writing, reading, and listening). (15 items)

2 Use of different classroom languages in a range of science class activities. (12 items)
3 Importance and difficulty of different English language skills in science classrooms. (17 items)
4 Teaching practices (11 items) and teaching beliefs (11 items) about the use of English in teaching science.
5 Language challenges (27 items) and coping strategies (11 items) in the EMI science classroom.

3.5.4 Student questionnaire

The same survey was administered to both groups of Grade 10 and 11 science students. The only differences were in the background information sought, such as the number of years of learning experience and the latest language and science achievements.

As with the teacher questionnaire, items were carefully adapted from previous studies with validated instruments after an intensive literature review. Items included classroom language use (Evans, 2000), student self-concepts and student belief in science (Yip et al., 2007a, b), and language challenges and strategies in EMI classrooms (Tatzl, 2011). The following items were covered in the student questionnaire:

1 Demographic data, self-reported science achievements, and English language abilities in four language skills (speaking, writing, reading, and listening). (15 items)
2 Use of different classroom languages in a range of science class activities. (12 items)
3 Importance and difficulty of different English language skills in science classrooms. (17 items)
4 Reflections on the impact of the medium of instruction on their science learning (3 items) and degree of importance and difficulty of English skills in science. (10 items)
5 Self-concepts (7 items) and beliefs about instructional activities (9 items) in learning science through English.
6 Language challenges (27 items) and coping strategies (11 items) in EMI science classrooms.

3.5.5 Development of the questionnaires

Both teacher and student questionnaires were designed to explore: (1) individual perceptions of EMI science learning, (2) language use in science, (3) English proficiency, (4) self-concepts, (5) teaching beliefs, (6) language challenges and strategies, and (7) relevant demographic information. All teachers and a small set of student participants were invited for follow-up interviews to obtain clarification or explanations about challenges and strategies identified in the surveys.

The questionnaire results informed my more guided questions for the second-round interviews with teachers and students in Stage 3.

For each surveyed item, the respondents were asked to indicate their preferences on a Likert scale. For example, students' self-concept was rated on a 5-point scale, with 1 indicating 'strongly disagree' and 5 indicating 'strongly agree'. Both questionnaires were developed in English. Only the student version was translated into Chinese and checked and proofread for accuracy by a professional Chinese–English translator. To ensure that students understood the questions, they were also provided with a bilingual version. The questionnaires were reviewed by three researchers experienced in science education, pedagogy, and linguistics to establish content validity in the categories of typical instructional activities in the science classroom and typical challenges and strategies for learning science through English. The researchers also ensured the construct validity of each survey item so that the measurements of each variable were consistent with the underlying reality. The procedures for analysing the questionnaire data and the selected statistical tests are further described in Section 3.7.3.

3.6   Data collection

The number of participants at each school and the data collected—video-recorded observations, interviews, and questionnaires—are summarised in Table 3.5.

3.6.1   *The teaching and learning materials of the participating schools*

It would have been ideal if the same topic had been taught across all observed classrooms to eliminate possible confounding variables, such as the type of teacher-led activities that could greatly affect the frequency of teacher–student

Table 3.5 Summary of main data collection

Schools	A, B, C, D	E, F, G, H	8
Classes	10 (5 Grade 10, 5 Grade 11)	15 (7 from Grade 10, 8 from Grade 11)	25
Subjects	6 biology, 2 physics, 2 chemistry	5 biology, 4 physics, 6 chemistry	-
Observed lessons	13	20	33
Teachers	6	13	19
Students (total)	346	450	796
Students (Grade 10)	158	210	368
Students (Grade 11)	188	240	428
	33 observed lessons in 30 visits		
	19 science teachers and 796 students were observed in the lesson observations		
	346 students from full EMI schools and 450 students from late–partial EMI schools		
	363 students from Grade 10 (3 classes from physics, 4 from chemistry, 5 from biology)		
	428 students from Grade 11 (3 classes from physics, 4 from chemistry, 6 from biology)		

interactions. However, this was an observational study that aimed to avoid alter-ing a teacher's normal teaching and students' learning behaviour. The science topic of the observed lessons was therefore not controlled. In the inclusion criteria, I already controlled for all participating schools to be following a similar teaching timeline. All of these schools were based on the same HKDSE science curriculum (CDC & HKEAA, 2007a, 2007b, 2007c) and teachers drew on this curriculum to develop their teaching plans. The curriculum outlines topics, teaching hours for each science topic, specific science knowledge expected for each topic, and the science learning activities (e.g., laboratory practices) students are expected to undertake in a specific timeline. From interviews with teachers and science panel heads, I learned that all eight participating schools followed the same teaching timeline, and delivered the same sets of science topics during the data collection period. All science teachers in the schools I visited claimed to follow the HKDSE curriculum when developing their teaching schedule, teacher-led activities, and laboratory classes.

All the science classes that I visited used textbooks as the primary source of teaching materials. Nearly all the participating schools used the same publisher for science textbooks. Some teachers used multimedia resources such as Power-Point slides and videos of laboratory experiments, which were developed by the textbook publishers and which the teachers projected onto a large screen in front of the class when teaching. The projections were the same set of learning materi-als that students could find in their textbooks. In Hong Kong, only a few publish-ers produce secondary science textbooks, and each publisher produces textbooks for only one specific science subject. In the schools I visited, textbooks from the publisher Aristo were used for chemistry; Oxford University Press textbooks were used for physics and biology.

3.6.2 *Pre-observations with teachers*

Pre-observation interviews were conducted with teachers briefly for two to three minutes at the beginning of each observed lesson to obtain information about the lesson. This included the teaching objectives of each lesson, teaching topics, prior knowledge, learning outcomes, types of teaching materials, and students' back-grounds. All teachers participated in the pre-observation interviews and these were audio-recorded in the teachers' native language (Cantonese) to allow them to express themselves freely.

All interviews with teachers and students were conducted in Cantonese, the participants' first language, to allow participants to easily recall what they thought about the observed lessons and to encourage them to provide comments on their perceptions and experiences of teaching and learning science in an English envi-ronment more generally.

3.6.3 *Video-recorded classroom observations*

A total of 33 lessons, mostly in double lesson sessions (40 minutes per lesson), were observed during 30 school visits. All lessons were video-recorded. Each class was observed twice or at least once in double periods. The researcher directly observed the class and took field notes about all teaching materials, class activities, participants' reactions regarding their classroom interactions, and the teachers' questioning skills.

3.6.4   *Procedures of video-recorded observation in each school*

Here is an example of an observation schedule for each school I visited. First, I met the school principal or science panel head to introduce the aims of the study, explain the research activities involved, negotiate the possible number of classroom visits, and discuss possible dates, times, and locations for classroom observation—in most cases, according to the teacher's schedule. During the first visit to each school, I had an informal meeting with the school principal, science subject heads, and science teachers who had expressed interest in participating. In the meeting, I answered their questions and addressed any concerns related to the study. Once I had obtained teacher and student consent, I arrived at the school at an agreed time.

The lessons observed varied according to the day of the week and the time of day (morning or afternoon). To prevent the teacher from altering his or her normal teaching practice, I did not explain in great detail what I would pay special attention to during my observation.

Sitting at the back of the classroom, I would observe and write detailed notes and comments on classroom interactions, questioning, and language issues. The teacher would be given a wireless microphone and the lesson would be video and audio-recorded with a camera at the back of the room. Video-recording would only take place with the consent of both teacher and students and if I had received no objection from any individual in the classroom. If a teacher or any of the students did not wish to be recorded, I would negotiate to audio record. If this was not acceptable, I would take only handwritten notes without recording, but this did not happen in any of the eight participating schools in my sample. I ended up observing each teacher for at least one double-period lesson in most schools I visited.

3.6.5   *Post-observation interviews with teachers*

After all the observations were completed, I invited each teacher to take part in a semi-structured interview, which was audio-recorded and lasted for around 30 minutes. All interviews were conducted in the participants' native language (Cantonese) and transcribed into English by the researcher. The interview was divided into two parts. In the first part, the participants were asked about their overall impression of the observed lessons, and their views about the pedagogical functions of the interactions in the lesson just observed. From time to time, I reminded participants of scenarios from the lesson that I had noted in my observation sheet. These scenarios were mostly related to class interactions and the teachers' questioning skills. I then invited the teachers to comment on their actions or student behaviour and to explain or justify the observed teaching behaviour. Examples of questions asked include 'Would you like to comment on why you asked such and such a question at that time?' or 'Why did you translate the word 'neutralisation' into Cantonese?'

In the second part of the interview, I asked for the teachers' views on using English to teach senior science classes, possible language challenges, and the individual strategies they used. The aim here was to find out their reflections on the usual practices in the lessons observed and to understand their views about the MOI practice and language use in English science classrooms. A few teachers shared views on the changes they had observed between the current six-year Hong Kong secondary curriculum and the previous seven-year secondary curriculum.

3.6.6   *Interviews with students*

Fifty-two students from the 8 participating schools were interviewed. Around 2–5 students for each class that had been observed were randomly selected by the researcher to be invited for a 10-minute focus group interview. In the pilot study, I tried one-on-one interviews with students and found that some students were reluctant to express their opinions. In the main study, I invited at least two students to be interviewed at the same time. This greatly encouraged students to give their opinions because they could explain their ideas by building on their peers' responses and providing additional examples. The focus group also allowed a genuine discussion among students, where students did not simply agree with one another but provided counterarguments of their own, supporting ideas on different issues raised in the interview. For example, I asked which language was their preferred MOI in learning science. Some students supported English, others supported Cantonese, and a few said code-switching was preferable.

All the interviews were conducted in Cantonese and audio-recorded without the presence of the teacher. The interview was designed to serve two purposes. Firstly, it was to check if the teacher and students behaved as usual in the observed lesson. Most of the students said that the observed lessons were normal and truly reflected their usual teaching and learning practices, thereby confirming the validity of the collected data. Secondly, the interview was an opportunity to explore student reflections on the observed classroom interactions and to elicit their views about learning science in the English medium and about individual language challenges and coping strategies. Table 3.6 summarises the interview data.

3.6.7   *Teacher and student questionnaires*

Once all observations had been carried out, the teachers and students who participated in the observations were invited to fill out a questionnaire. Respondents

Table 3.6 Summary of the interview data

	Early–full EMI schools	Late–partial EMI schools	Total
No. of pre- and post-interviews	12	22	34
No. of teachers	6 (5 male, 1 female)	11 (10 male, 1 female)	17
Subjects	3 biology, 2 physics, 1 chemistry	4 biology, 4 physics, 5 chemistry	—
No. of student focus group interviews	8 (Grade 10:4; Grade 11:4)	14 (Grade 10:7; Grade 11:7)	22
No. of students	20 (Grade 10:10; Grade 11:10)	32 (Grade 10:17; Grade 11:15)	52
Total	19 teachers were interviewed; 6 teachers from the EMI schools; and 13 teachers from the MOI-switching schools. In terms of subject distribution, there were 6 physics teachers, 6 chemistry teachers, and 7 biology teachers.		
	43 teacher interviews were conducted, including 25 pre-observation interviews (2–3 minutes) and 19 post-observation interviews (15–20 minutes)		
	52 students were interviewed after observation. 19 students from the EMI schools, and 33 students from the MOI-switching schools. 27 Grade 10 students and 25 Grade 11 students. 12 physics students, 16 chemistry students, and 24 biology students		
	22 student interviews were conducted. 11 groups from Grade 10 and 11 groups from Grade 11, each group consisting of 2–5 students. (10–15 minutes)		

only completed the questionnaires after all observations had finished. The researcher was present when the students completed the questionnaires to avoid any possible influence from the teacher. Students were provided with sufficient time to complete the questionnaire and they were allowed to ask questions to ensure they clearly understood all the instructions.

Most teachers and students could understand and follow all the questions in the questionnaires. Most classes finished the questionnaire in about 15–20 minutes. Overall, 19 teachers and 796 students from the eight participating schools filled in the questionnaires—only 545 students returned their questionnaires to the researcher, meaning the return rate for student questionnaires was 68%. All returned questionnaires were checked carefully to ensure every item was filled in. If there were any missing pages or items, the questionnaires would be returned to the school and collected after the items were completed. The questionnaires of 19 teachers, 368 Grade 10 students, and 428 Grade 11 students were collected. Among the collected student questionnaires, 346 were from the early–full EMI schools and 450 were from the late–partial EMI schools. Table 3.7 summarises the demographic backgrounds of the student respondents to the questionnaires, and Table 3.8 summarises the demographic backgrounds of the teacher respondents.

Table 3.7 Demographic backgrounds of the student respondents to the questionnaire

Demographic characteristics	Student returned questionnaires (n = 545); response rate= 90%	
School types	Early–full EMI (*n* = 284)	Late–partial EMI (*n* = 261)
Sex		
Male	101 (19%)	136 (25%)
Female	182 (33%)	125 (23%)
Age		
15	0	40 (7%)
16	56 (10%)	102 (19%)
17	150 (28%)	88 (16%)
18	59 (11%)	24 (4%)
19 or above	19 (3%)	7 (1%)
First language		
Cantonese	282 (52%)	258 (47%)
Mandarin	2 (0.4%)	3 (1%)
Place of primary education		
Hong Kong	282 (52%)	258 (47%)
Mainland China	2 (0.4%)	3 (1%)

Table 3.8 Demographic backgrounds of teacher respondents to the questionnaire

Demographic characteristics of respondents	Teachers (n = 19); no. (overall 100%)	
School types	Early–full EMI (*n* = 6)	Late–partial EMI (*n* = 13)
Sex		
Male	5 (26%);	11 (58%)
Female	1 (5%)	2 (11%)
Age:		
20–30 years	3 (16%)	0
31–40	2 (11%)	5 (26%)
41–50	1 (5%)	4 (21%)
51–60	0	4 (21%)
Teaching areas:		
Physics	2 (11%)	4 (21%)
Chemistry	1 (5%)	5 (26%)
Biology	2 (11%)	4 (21%)
Years of teaching experience	Mean (range, median)	Mean (range, median)
at secondary school	10.2 (2–22, 9)	20.8 (6–35, 19)
in English medium of instruction	8.3 (4–15, 5.5)	18.1 (3–40, 17)
in junior levels (Grades 7–9)	7.5 (4–15, 5.5)	10.1 (0–35, 3)
in senior levels (Grades 10–12)	10.0 (1–22, 9)	20.3 (2–35, 19)
in HKDSE (G10–12) (since 2010)	4.2 (1–6, 5)	4.1 (3–6, 4)

(Continued)

Table 3.8 (Continued)

Demographic characteristics of respondents	Teachers (n = 19); no. (overall 100%)	
in Cert. Level (G10–11) (until 2010)	5.8 (0–18, 4)	12.5 (2–32, 13)
in A-Level (G12–13) (until 2010)	5.8 (0–18, 4)	15.1 (0–30, 16)
Self-reported English Language skills[a]	Mean (range, median)	Mean (range, median)
Listening	3.8 (3–4, 4)	3.4 (2–4, 3)
Speaking	3.7 (3–4, 4)	3.2 (2–4, 3)
Writing	3.5 (3–4, 3.5)	3.2 (2–4, 3)
Reading	3.8 (3–4, 4)	3.5 (3–4, 3)
Received EMI training?	Yes: 2; No: 6	Yes: 0; No: 13

[a]   Scale=1 (poor), 2 (fair), 3 (moderate), 4 (good) and 5 (excellent).

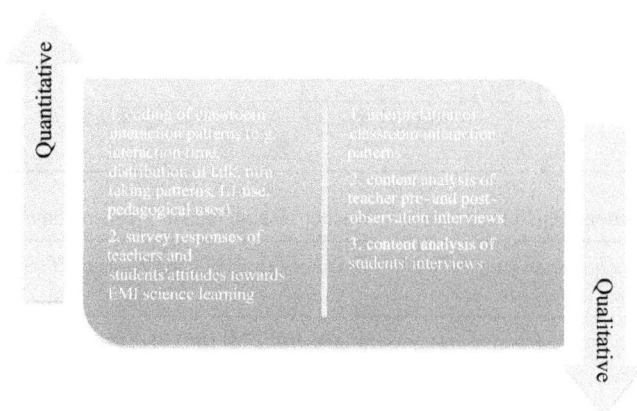

Figure 3.6 Research approaches to analysing multiple datasets.

3.7   Analysis of the data

The following sections briefly discuss how I analysed the qualitative and quantitative data from multiple sources (see Figure 3.6).

3.7.1   *Analysing classroom interaction in a quantitative manner*

The procedures for analysing teacher and student talk were based on the methods of Lo and Macaro (2012), with specific attention to the teacher–student interactions, the teachers' questioning types and the teachers' explanations that are related to the aims of the study. The observed classroom interactions were handled in quantitative and qualitative manners. Figure 3.7 summarises the key steps for analysing teacher and student talk.

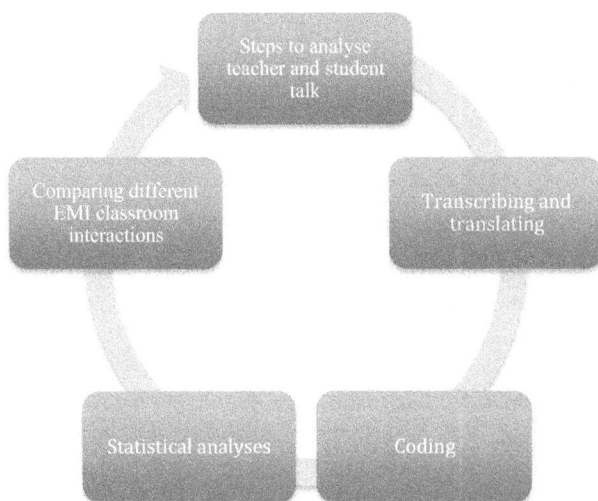

Figure 3.7 Key steps for analysing teacher and student talk.

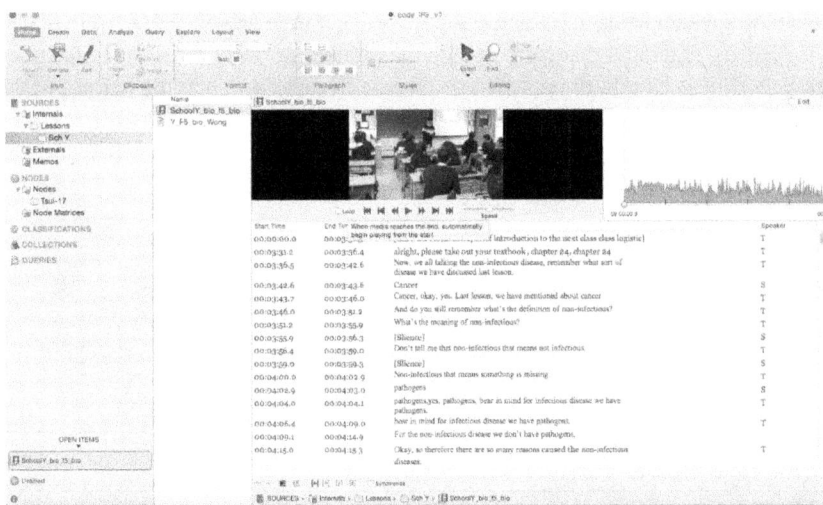

Figure 3.8 The *NVivo* programme window showing the transcribing process.

3.7.2 *Transcribing and translating teacher and student talk*

The 33 video-recorded science lessons (physics, chemistry, and biology) from four full EMI and four late–partial EMI schools were transcribed using the *NVivo* software package (see Figure 3.8). The data was transcribed in English and analysed, together with the unstructured observation notes about the teaching procedures and the researcher's comments. As observed lessons were mainly teacher-centred, only teacher–student interactions and some student–student interactions in whole-class situations were transcribed. Student–student interactions during group work or private conversations that were captured by the recorders were excluded. The lessons were first transcribed into English and supplemented with Chinese transcriptions at points when participants were using their first language or code-switching.

3.7.3   Coding for teacher and student talk

An initial coding system was adopted from Lo and Macaro's (2012) study, with specific attention paid to exploring classroom interactions in the science classroom. The coding framework aims to analyse classroom discourse in a quantifiable manner. After the lessons were transcribed and translated, these transcription files and the corresponding video files were loaded into the *NVivo* software package for further coding. First, the teachers' and students' turns were defined, identifying different 'acts'.[7] These were then categorised into IRF exchange sequences based on established analytical frameworks (Sinclair & Coulthard, 1975; Tsui, 1985). In the *NVivo* programme window (see Figure 3.9), the left-hand side shows the hierarchy of different 'acts' and the IRF sequences. The right-hand side shows the length (in seconds) of each act in the video-recorded lesson and the actual talk (what the speaker said) in each act. After coding all the acts in one lesson, an output file was generated that summarised the number of occurrences of each coded 'act', their time length, the percentage of time of each 'act', and the pattern of the overall IRF sequence in one observed lesson.

With all 33 lessons coded, all the teacher and student talk were categorised. The information generated from *NVivo* helped me calculate the following eight items:

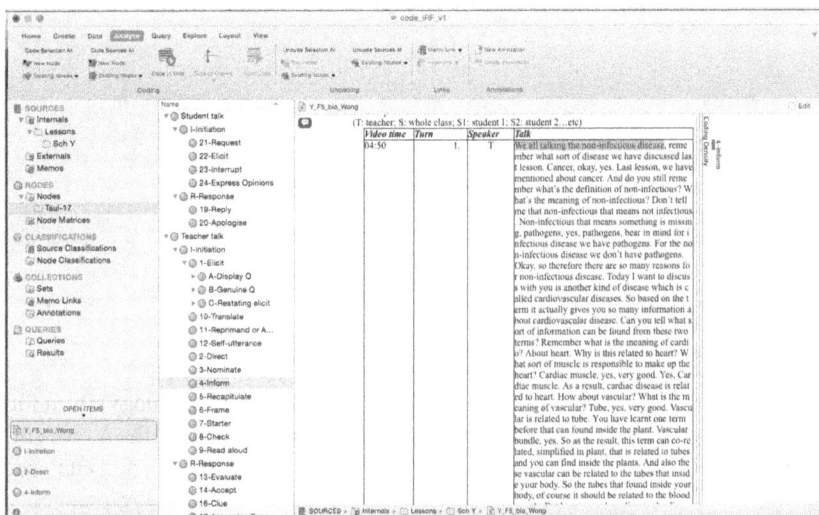

Figure 3.9 The *NVivo* programme window showing the coding process.

1 Overall time interacting within a lesson.
2 Proportion and length of teacher and student talk.
3 Number of Initiation-Response-Feedback (IRF) exchange sequence patterns between teachers and students.
4 The pedagogical functions of teacher and student talk.
5 Number of L1 and L2 use.
6 L1 use.
7 Science teachers' questions.
8 Science teachers' feedback.

The coding categories summarised in Figure 3.10 were then used to compare patterns of interactions in the science classrooms of early–full and late–partial EMI science classrooms and of Grade 10 and 11 in each group. Such analysis addressed the extent to which classroom interactions were similar or different between different groups (RQ2 and RQ3). Table 3.9 summarises each of the definitions and subcategories. This coding frame provides the key elements for analysing teacher and student talk quantitatively (Table 3.9).

Aspect of Analysis	
1. Teaching purposes *(Mortimer and Scott, 2003)*	*a) Opening up the problem*
	b) Exploring and working on students' views; introducing and developing the scientific story
	c) Guiding students to work with scientific ideas and supporting internalisation
	d) Guiding students to apply, and expand on the use of, the scientific view, and handing over responsibility for its use
	e) Maintaining the development of the scientific story
2. Content *(Mortimer and Scott, 2003)*	*a) Everyday-specific*
	b) Description, explanation, generalisation
	c) Empirical-theoretical
3. Communicative approach *(Mortimer and Scott, 2003)*	*a) Interactive/ non-interactive*
	b) Authoritative/ dialogic
4. Patterns of interaction *(Mortimer and Scott, 2003)*	*a) IRF*
	b) extended IRF
5a. Forms of interaction	*a) IRF/E (Initiation-Response-Feedback/Evaluation)*
	b) IRFRF (Initiation-Response- Feedback- Response- Feedback)
5b. Functions of interaction	The seventeen categories of acts in classifying second-language classroom talk (Tsui, 1985)
	Chin's (2006) Taxonomy of teacher's feedbacks to students' responses in science classroom

Figure 3.10 The coding framework for teacher and student talk.

6. **Teaching interventions** **(Mortimer and Scott, 2003)**	*a) Shaping ideas* *b) Selecting ideas* *c) Marking key ideas* *d) Sharing ideas* *e) Checking student understanding* *f) Reviewing*
7. **Teacher's question types** **(Yip 2004)**	*1) Recalling* *2) Explanation* *3) Analysis* *4) Evaluation* *5) Synthesis* *6) Motivation* *7) Eliciting* *8) Challenging* *9) Extending* *10) Application*
8. **The choice of instructional language (L1/L2), proportion of teacher/ student talk**	*Timed analysis (Macaro, 2001)* *— sampling the interactions every five seconds and then code according to speaker and the language being spoken*
9. **The distribution and allocation of teacher/ student turns**	*Turn analysis (Macaro & Mutton, 2002)* *— calculating the mean length of each speaker's turn)*
10. **Type of code-switching patterns; (Myers-Scotton, 1988)**	*a) Intra-sentential (within a sentence)* *b) Inter-sentential (between sentences)*

Figure 3.10 (Continued).

3.7.4 Statistical analyses for comparing different groups of classroom interactions

After the basic information about the interaction patterns between Grades 10 and 11 was obtained in each type of school, further statistical analyses were performed. The purpose was to estimate the extent to which interaction patterns in each group were significantly different between early–full and late–partial EMI schools.

A number of independent sample *t*-tests (for parametric data) or Mann–Whitney tests (for non-parametric data) were performed to estimate the observed differences in classroom interactions between Grades 10 and 11 and between early–full and late–partial EMI schools. First, the normality of the distribution of data was examined. The statistical procedure includes calculating the values of 'skewness' and 'kurtosis' of the data, and these numbers should fall into the range of -2.5 to 2.5 for the data to be regarded as normally distributed. Second, the appropriate statistical test was applied to the data set. The *t*-test would be used if the data was

Table 3.9 The framework for coding the classroom interactions

Interactional variable		Dimension analysed
a. The interaction time	i. Interaction time	Total recorded lesson time, total interaction time, and percentages of interaction time
	ii. Proportion of talk	Proportion of teacher talk, students talk, and percentages of talk for each observed lesson
	iii. Length of turns	Turn analysis: calculating the mean length of each speaker's turn
	iv. Ratio between L1 and L2 use	Proportion of teacher and student talk in L1 (Cantonese) and L2 (English)
b. The pedagogical functions of interaction		17 pedagogical functions of IRF patterns classifying classroom talk between teacher and students (Tsui, 1985)
c. The science teacher questions		Science teacher questions in developing student's science understanding (Yip, 2004)
d. The science teacher's feedback		Classification of science teachers' feedback to students' responses in IRF patterns (Chin, 2006)
e. The use of L1	Type of code-switching patterns	a) Intra-sentential (within a sentence); b) Inter-sentential (between sentences)

normally distributed and the Mann–Whitney test would be used if the data was not normally distributed. Third, after applying the appropriate statistical test, the mean of the two groups was then compared, and the results of the differences such as the proportion of the teachers' and students' talks, the average time length of the teachers' and students' turns, and the IRF sequences were reported.

3.8 Analysing classroom interactions in a qualitative manner

The quantified analysis outlined above generated numeric accounts of the classroom interactions observed in each group and provided comparable results between early–full and late–partial EMI schools. In a separate analytical phase, I analysed classroom interaction qualitatively, to provide a comprehensive picture of the actual teaching and learning process in the observed classrooms. In analysing the classroom interactions, I am interested in what the science teachers did in the teaching process and the functions of their turns at talk (science pedagogy), what kind of languages the teacher and students used for interaction (the language of science), and how teachers and students interacted to construct content knowledge (thematic understanding of science concepts).

This phase of analysis consists of three major components: (1) analysing science pedagogy, (2) analysing the language of science, and (3) analysing how students develop a thematic understanding of science concepts through teacher–student

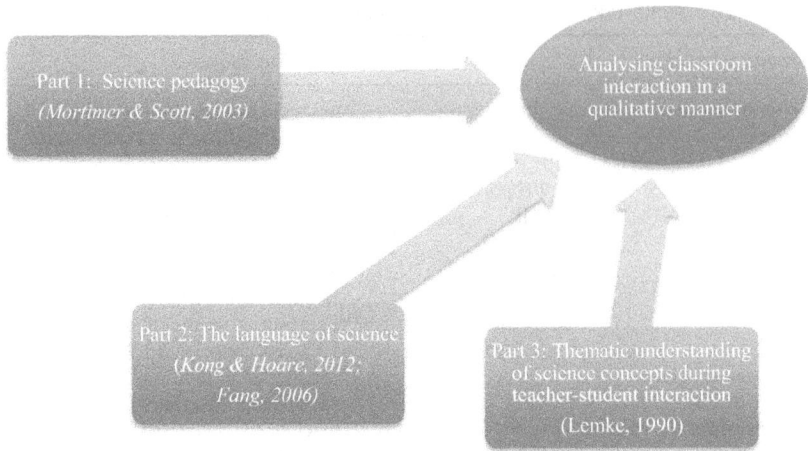

Figure 3.11 Framework used to analyse classroom interactions qualitatively.

interaction. This combined framework for analysing the language of science in the classroom interactions observed is summarised in Figure 3.11. These analyses help to address RQ1, i.e. 'What is the nature of the classroom interaction in an EMI science classroom?' The diagram in Figure 3.11 illustrates the framework used to analyse classroom interactions qualitatively.

3.8.1   Science pedagogy

In the qualitative analysis of classroom interaction, I examined the science pedagogy using a frequently used framework (Mortimer & Scott, 2003). Table 3.11 shows the proposed coding frame for analysing science pedagogy. In this analytical framework, specific attention is paid to the following aspects of analysis: (1) teaching, (2) content, (3) communicative approach, (4) patterns of interaction, (5) functions and forms of interaction, and (6) teaching interventions.

Table 3.10 The proposed coding framework for analysing science pedagogy

Science Pedagogy (Mortimer and Scott, 2003)	
a. Teaching purposes	Teaching purposes in relation to the science being taught, at each phase of the observed lesson
b. Content	Content of classroom interactions
c. Communicative approach	Ways in which a teacher guides students in addressing science ideas developed in a lesson through frequent or limited interactions
d. Patterns of interaction	Distinctive patterns of interactions observed between teacher and students: IRF vs extended IRF
e. Teaching interventions	Ways in which the teacher intervenes in order to develop the scientific story by addressing students' scientific ideas

3.8.2 *The language of science*

One of the relevant features in teacher and student talk in an EMI context is the use of academic language. Academic language plays a critical role in interactions between teachers and students as they construct science experiences and shapes how teachers and students explain abstract scientific phenomena and express their scientific ideas. In this part of the analysis, I adapted Kong and Hoare's (2012) framework of the characteristics of academic language in science classrooms, as reviewed in Chapter 2. This framework can serve as a tool for analysing the complexity of academic language used in teacher and student talk. It also explains how teachers can develop students' academic language through interactions in an EMI science classroom.

To provide a comprehensive understanding of the language features used in a science classroom, I adopted a framework that describes the linguistic features of school science in addition to Kong and Hoare's framework. Fang (2006) examined US high school students' language use in content-based instruction science classrooms. His framework explains the language of science that science students experience in an English as a second language (ESL) classroom. It also describes language demands and challenges in the ESL context. This framework usefully supplements Kong and Hoare's framework in providing a more extensive list of linguistic features in science and possible language challenges for EMI science students. Table 3.11 summarises the two frameworks used to describe the language of science.

Table 3.11 The frameworks adapted to describe the language of science

Kong and Hoare (2012)	Fang (2006)
Framework describing the language of science in teacher and student talk	Framework describing the language features of science
Use of subject-specific vocabulary	Technical vocabulary
Use of nominalisations (normally used as grammatical metaphors, when nominalised words can be unpacked)	Ordinary words with multiple meanings
Use of grammatical metaphors (when meaning is expressed in uncommon grammatical forms)	Words with multiple grammatical functions
Use of complex noun phrases (noun phrases with a head plus pre-modification and or post-modification)	Grammatical words Ellipsis Subordinate clauses prepositional phrases
High lexical density (measured by the number of content words divided by the total number of words; content words include lexical items, such as nouns, adjectives, verbs, adverbs, and functional words include grammatical items, such as prepositions, pronouns, auxiliary verbs, conjunctions, demonstratives)	Abstract nouns Lengthy nouns Complex sentences Interruption construction Passive voice
Use of the language of logical meaning or knowledge relationships (such as classification, definition, cause-and-effect)	

3.8.3   Developing students' thematic understanding through extended IRF exchanges between teachers and students

Lemke's (1990) concept of thematic development is used to examine how students develop an understanding of science concepts through classroom interactions. This part of the analysis focused principally on how science teachers modify their talk to develop students' understanding of science through interaction. To do this, in each IRF sequence, I analysed how teachers responded to students by either selecting the students' science ideas in their responses or by modifying the students' answers to formulate further questions in their teacher feedback. This prompted a meaningful interaction that developed students' thematic understanding of science topics in terms of taxonomic knowledge of related science vocabulary.

I therefore analysed thematic development in teachers' talk, focusing on the selection and modification of students' answers in developing a coherent understanding of science. Linguistically, the analysis focused on whether and how teachers paid attention to explaining (1) the semantic relations between words and nominal phrases and (2) the logical relations between phrases.

Through the analysis of IRF triadic exchanges, I aimed to see to what extent teachers in these two types of EMI schools are skilful enough to make the science content more accessible to students in L2. The teachers' selection of students' science ideas in their responses and the teachers' modified input are key aspects in producing input that is comprehensible to students and that develops their science understanding and familiarity with the language of school science.

To achieve this goal, I examined IRF sequences in episodes where teachers were teaching a science concept, particularly when teachers asked a question and provided feedback to students. I paid particular attention to the following dimensions of teacher input:

1 How the teacher selected students' ideas:
    a When students answered teacher questions, what aspects of students' ideas were picked up on by the teacher in subsequent turns by the teacher?
    b What kinds of ideas from the students' talk were chosen by the teacher to become the focus of later exchanges?
2 How the teacher modified his/her talk:
    a What kind of feedback did the teacher give?
    b Did the teacher modify the feedback based on the students' responses?
    c What decision(s) did the teacher make in reacting to students' responses (i.e., Did teachers ignore, acknowledge, or direct students to the teachers' intended specific view on the scientific phenomena, or expand on the student's idea)?

This aspect of the analysis is in line with how others have analysed science pedagogy. For example, Mortimer and Scott (2003) analyse the four communicative approaches in the science classroom (dialogic vs. authoritative) and teachers' interventions (ways that teachers intervene with students to develop the scientific story and make it accessible to all the students in the lesson).

To conclude, I have summarised several frameworks for examining classroom interaction qualitatively. I have explained each of the categories and their relevance in providing an integrated framework for analysis.

3.9 Analysing interview data

All the data from teacher and student interviews was audio-recorded, transcribed with pseudonyms, and saved as a text file in the *NVivo* software. The interview data was treated with a similar procedure to that of the observation data in *NVivo*. A two-step coding process was used for an in-depth qualitative analysis of the transcribed interview data (Corbin and Strauss, 2008). First, each interview transcript was imported into the software and an initial open coding was developed for each pattern or recurrent theme related to the main focus of the interview questions (i.e., views on EMI, language challenges, strategies). Then, specific parts of the text in these segments were coded. Called 'axial coding', this is a process of relating categories to their subcategories, again to represent sub-themes to which the statements belong.

Using a grounded theory approach, the final version of the coding system was gradually developed with major themes drawn from the initial coding list. The final coding list with emerging findings from the major themes in the interview data was used to refine or amend the items covered in the larger-scale quantitative questionnaire in stages two and three. The coded interview transcripts were carefully read and repeatedly compared, sorted, recoded, and explored for connections among coded segments. About 10% of the data (observations, interviews) were independently coded by two trained raters according to the coding sheet developed, to ensure the reliability of the coding framework. The two ratings establish an inter-rater reliability of $(k > 0.8)$ using Cohen's Kappa coefficient (Robson, 2002).

3.10 Quantitative data: teacher and student questionnaires

The questionnaire data were manually entered into the SPSS software and descriptive statistical reports were generated for both the teacher and student groups. Means and standard errors were calculated. Inferential statistics (independent *t*-tests or Mann–Whitney tests) were used to calculate mean scores for subgroups (Grade 10 vs. Grade 11, early–full EMI schools vs. late–partial EMI schools) and compared statistically. Inferential statistics, including those groups that are statistically significant, were reported.

The first of the two methods of analysis concerns teacher and student views about particular issues—teaching practices and beliefs about English in science and students' self-concepts and beliefs, respectively. These items can be described using summary measures such as the mean and standard deviation of responses for each item.

The second approach involves the underlying factors that determine teachers' and students' overt responses to individual language challenges and strategies. Exploratory factors analysis was used to calculate each factor loading in each item (i.e., challenges and strategies). Then, the number of factors was decided by admitting factors with eigenvalues greater than 1. Principal component factor analysis was conducted to investigate the underlying constructs and identify the major challenges in classroom instruction. Cronbach's *alpha* values were calculated to check the internal consistency. This ensured that the degree of each indicator variable measuring the same latent construct was correlated (Field, 2013). Factors explaining language challenges and coping strategies can be interpreted as important indicators of the quality of classroom interactions.

For Grade 11 students, self-reported proficiency in English and science achievements were collected via a questionnaire survey. After controlling for prior MOI background, school type, and academic achievement before secondary school, the two scores for each student were entered into SPSS and either the Pearson product moments (if the data were normally distributed) or the Spearman rho correlation coefficients (for non-parametric data) were run. This determined if there was a relationship between students' English language proficiency and their Grade 11 science academic performance.

3.11 Summary

This chapter has presented the different stages of research design, research questions, and data collection. I have described the study population, the sampling frame, the criteria for selecting participants, and the data collection process. To illustrate how the research questions were addressed, the data analysis framework and procedures were also presented. The data collected for this study consisted of 33 observation videos, pre- and post-observation interviews with 19 teachers, focus group interviews with 52 students, 19 teacher questionnaires, and 545 student questionnaires. The next chapter will provide an analysis of classroom interactions in different EMI science classrooms. I will quantitatively analyse classroom interaction by comparing (i) early–full and late–partial EMI schools, and (ii) Grades 10 and 11 for each school type.

Notes

1 Within the 33 observations, there are 10 lessons in full EMI schools and 23 lessons in late–partial EMI schools; 15 Grade 10 lessons and 18 Grade 11 lessons; 8 physics lessons, 10 chemistry lessons, and 15 biology lessons.
2 Source: Education Bureau, Hong Kong Special Administrative Region, Figures and Statistics on Secondary Education, http://www.edb.gov.hk/tc/about-edb/publications-stat/figures/sec.html.
3 Primary Six school-leavers in Hong Kong are rated on a three-band scale according to their academic performance, or the AAI score (Academic Aptitude Index), to determine which secondary schools they will be admitted to. The AAI score is based on a student's academic achievements in the school-based examinations in primary five and six. Band

1 is the above-average band in the three-tier categorisation system, with high academic achievement. Band 2 is average academic achievement. Band 3 is below the average academic standard.

4 To avoid noticeable demographic differences such as those in residents' socioeconomic status and educational qualifications across different areas in Hong Kong, only schools located in districts with similar demographic characteristics were selected and areas where mainly lower- and middle-class residents were preferred. For students' academic ability, student academic achievement before admission and performance in public examinations of the schools were also considered. In Hong Kong, students with better academic results usually enrol in full EMI schools, which are considered more prestigious than CMI or late–partial EMI schools and promise improved prospects for students. This implies that academic performance in most full EMI schools is generally better than that of late–partial EMI schools. For a comprehensive view of the situation of EMI instruction in different types of schools, 4 full EMI and 4 late–partial EMI schools with comparable academic ability of the student intake were included. The CMI schools were not chosen because the academic performance of the student intake was significantly different from that of the early–full EMI and late–partial EMI schools and because all teaching and learning (activities, interactions, assessments, and teaching materials) were in Chinese. Most of these CMI schools did not change their MOI at senior levels, suggesting limited English proficiency among their students.

5 The number of students in the observed lessons was informed by the participating teachers and confirmed with the actual counting by the researcher during the observation.

6 Most of the students who migrated to Hong Kong were from the southern cities of Mainland China, e.g. Guangdong province, where they are regarded as Cantonese and speak Cantonese as their first language and Mandarin as their official language in school.

7 According to Sinclair and Coulthard (1975, p. 27), the units at the lowest rank of discourse are *acts* that carry performative functions.

References

CDC & HKEAA. (2007a). *Science education key learning area biology curriculum and assessment guide (Secondary 4–6)*. Hong Kong.

CDC & HKEAA. (2007b). *Science education key learning area chemistry curriculum and assessment guide (Secondary 4–6)*. Hong Kong.

CDC & HKEAA. (2007c). *Science education key learning area physics curriculum and assessment guide* (Vol. 2007). Hong Kong.

Census and Statistics Department. (2014). Retrieved from http://www.bycensus2006.gov.hk/en/data/data2/index.htm

Chin, C. (2006). Classroom interaction in science: Teacher questioning and feedback to students' responses. *International Journal of Science Education, 28*(11), 1315–1346. http://doi.org/10.1080/09500690600621100

Committee on Home-School Co-operation. (2014). The current version of secondary school profiles. Retrieved from http://www.chsc.hk/ssp2013/chi/index.php

Corbin, J. H., & Strauss, A. L. (Ed.). (2008). *Basics of qualitative research: Techniques and procedures for developing grounded theory* (3rd ed). Thousand Oaks, CA: Sage Publications.

Creswell, J. W., & Plano Clark, V. L. (2011). *Designing and conducting mixed methods research* (2nd ed.). Los Angeles, CA: SAGE Publications.

Curriculum Development Council (CDC). (2002). *Science education KLA curriculum guide (P1-S3)*. Hong Kong.

Dörnyei, Z. (2007). *Research methods in applied linguistics.* New York: Oxford University Press.

Evans, S. (2000). Classroom language use in Hong Kong's English-medium secondary schools. *Educational Research Journal, 15*(1), 19–43.

Fang, Z. (2006). The language demands of science reading in middle school. *International Journal of Science Education, 28*(5), 491–520. http://doi.org/10.1080/09500690500339092

Field, A. (2013). *Discovering statistics using IBM SPSS statistics.* Los Angeles, London, New Delhi: Sage.

Hoare, P. (2003). *Effective teaching of science through English in Hong Kong secondary schools.* https://doi.org/10.5353/th_b2976829

Kington, A., Sammons, P., Day, C., & Regan, E. (2011). Stories and statistics: Describing a mixed methods study of effective classroom practice. *Journal of Mixed Methods Research, 5*(2), 103–125. http://doi.org/10.1177/1558689810396092

Legislative Council. (2009). *Fine-tuning the medium of instruction for secondary schools* (Vol. 9). Hong Kong. Retrieved from http://sc.legco.gov.hk/sc/www.legco.gov.hk/yr08-09/english/panels/ed/papers/ed0115cb2-623-1-e.pdf

Lemke, J. L. (1990). *Talking science: Language, learning and values.* Norwood, New Jersey: Ablex Publishing Corporation.

Lo, Y. Y., & Macaro, E. (2012). The medium of instruction and classroom interaction: Evidence from Hong Kong secondary schools. *International Journal of Bilingual Education and Bilingualism, 15*(1), 29–52. http://doi.org/10.1080/13670050.2011.588307

Lo, Y. Y., & Lo, E. S. C. (2014). A meta-analysis of the effectiveness of English-medium education in Hong Kong. *Review of Educational Research, 84*(1), 47–73. http://doi.org/10.3102/0034654313499615

Mortimer, E., & Scott, P. (2003). *Meaning making in secondary science classrooms.* Berkshire, England: Open University Press.

Patton, M. Q. (1999). Enhancing the quality and credibility of qualitative analysis. *Health Services Research, 34,* 1189–1208.

Robson, C. (2002). *Real world research: A resource for social scientists and practitioner-researchers.* (2nd ed.). Oxford: Blackwell.

Sinclair, J., & Coulthard, M. (1975). *Towards an analysis of discourse.* Oxford: Oxford University Press.

Tatzl, D. (2011). English-medium masters' programmes at an Austrian university of applied sciences: Attitudes, experiences and challenges. *Journal of English for Academic Purposes, 10*(4), 252–270. http://doi.org/10.1016/j.jeap.2011.08.003

Tsui, A. B. M. (1985). Analysing input and interaction in second language classrooms. *RELC Journal, 16*(1), 8–32.

Wong, K. H. (2012). *Implementation of problem-based learning in junior secondary science curriculum.* University of Hong Kong. Retrieved from http://hdl.handle.net/10722/193087

Yip, D. Y. (2004). Questioning skills for conceptual change in science instruction. *Journal of Biological Education, 38*(2), 76–83. http://doi.org/10.1080/00219266.2004.9655905

Yip, D. Y., Coyle, D., & Tsang, W. K. (2007a). Evaluation of the effects of the medium of instruction on science learning of Hong Kong secondary students: Instructional activities in science lessons. *Education Journal, 35*(2), 77–107.

Yip, D. Y., Coyle, D., & Tsang, W. K. (2007b). Medium of instruction on science students: Instructional activities in science lessons. *Education Journal, 35*(2), 77–107.

4 Quantitative analysis of classroom interactions in early–full and late–partial EMI science classrooms

This chapter will present the analysis of classroom interactions in varying EMI science classroom settings. In this analysis of classroom interactions, a quantitative approach was first used in comparing (i) early–full and late–partial EMI schools and (ii) Grades 10 and 11 in each school type. In total, 33 lessons were video-recorded, and 31 hours and 45 minutes of classroom interactions were transcribed and analysed using the *NVivo* software package. Around 22.5 hours of interactions between teachers and students were analysed, 8.5 hours of which were pauses and non-interactions. The observations were transcribed and checked by a bilingual speaker for accuracy, and the interactions were codified according to the integrated framework illustrated in Chapter 3.

The first part of the following analysis compares the similarities and differences in interactional patterns quantitatively between the two types of EMI schools. It includes seven aspects of analysis in terms of classroom interactions:

1 interaction time
2 proportion and length of teacher and student talk
3 number of initiations and responses
4 wait time between teacher and student turns
5 proportion of teacher and student talk in L1 and L2, and L1 patterns
6 pedagogical functions of teacher and student talk
7 teachers' questions.

These analyses provide an objective measure of the differences and similarities according to the pedagogical functions. To understand how the teachers and students interacted differently in these schools and the purposes of the interactions, I also coded interactions using frameworks drawn from previous studies.

The second part of the analysis examines three aspects: (1) the pedagogical functions of teacher and student talk, (2) teachers' questions and (3) teachers' feedback. As a whole, the findings address two of the book's five research questions:

1 What are the differences and similarities in classroom interaction between early–full and late–partial EMI schools?
2 What are the differences and similarities in classroom interaction between Grades 10 and 11 in each group?

DOI: 10.4324/9781003001454-5

4.1   Data codification

The coding framework for analysing the classroom interactions in both types of schools was adopted from Lo and Macaro's (2012) mixed-method study, with specific attention paid to classroom interactions in senior secondary science classes. As such, the data codified was based on the following aspects of analysis:

1   interaction time
2   proportion and length of teacher and student talk
3   number of initiations and responses
4   wait time
5   proportion of L1 and L2 use
6   L1 patterns
7   types of teacher feedback
8   the pedagogical functions of teacher and student talk
9   science teachers' questions
10   science teachers' feedback.

Excerpts from the recorded transcripts are also included to further illustrate how the teachers and students interacted in early–full and late–partial EMI science classrooms.

4.2   The quantitative analysis of classroom interactions between early–full and late–partial EMI science classrooms

4.2.1   *Checking normality*

For inferential statistics, the assumptions of parametric tests were checked for the data. The interaction time, teacher and student talk, and length of teacher and student turn in all of the lessons were calculated as percentages (%) and inserted into the SPSS software package.

The coding covered 12 variables: percentages of interaction time, percentages of teacher talk, percentages of student talk, percentages of average length of each teacher turn, percentage of length of each student turn, percentages of teacher feedback, percentage of wait time, percentages of L1 use and L2 use, percentages of inter- and intra-sentential code-switching, percentage of pedagogical functions, percentage of teachers' questions, and percentage of science teacher questions.

All variables were screened for normality through histograms and Q–Q plots. Kolmogorov–Smirnov tests were also conducted to calculate the values of skewness and kurtosis for each variable to check their normality. All the variables were not normally distributed and thus, the Mann–Whitney U test was used along with the Bonferroni correction. Cohen's (1988) criterion was also used for judging the effect size of certain variables.

4.2.2 Percentage of interaction time

Table 4.1 shows the total average lesson length for a total of 32 lessons, the average lesson length,[1] the average interaction time per lesson,[2] and the percentage of interaction time[3] between school types and between grades in each school group, while the Mann–Whitney U test was used to compare the percentage of interaction between school types and between grades in each school group.

The percentages of interaction time between the early–full EMI schools (69%) and the late–partial EMI (73%) were similar. In both types of EMI schools, the percentages of interaction time in Grade 11 were lower than in Grade 10. Both types of schools had 9% less interaction time in Grade 11 than in Grade 10.

For the comparison between school types, the Mann–Whitney U test revealed no significant difference in the percentages of interaction time of early–full EMI schools (md = 86.75%, n = 10) and late–partial EMI schools (md = 83%, n = 23), U = 102, z = −5.09, p = 0.611, r = 0.089.[4]

For the comparison between grades in each group, the percentages of interaction time in the early–full EMI schools did not differ significantly between Grade 10 (md = 84.4%, n = 5) and Grade 11 (md = 89.1%, n = 5), U = 11, z = −0.313, p = 0.754, r = −0.099. In the late–partial EMI schools, the percentages of interaction time also did not differ significantly between Grade 10 (md = 82.75%, n = 12) and Grade 11 (md = 83%, n = 11), U = 59, z = −0.431, p = 0.667, r = −0.090.

Thus, the Mann–Whitney U test revealed that there was no statistically significant difference between (a) the early–full and late–partial EMI schools or (b) between Grades 10 and 11 in each group, regarding their percentages of interaction time.

Table 4.1 Mann–Whitney U test on the percentages of interaction time by school type and by grade

	School	Grade	n (lessons)	Average lesson length (mins)	Average interaction time per lesson (mins)	Percentages of interaction time	Median (SD)	Asymp. Sig. (two-tailed)
Between school types	Early–full		10	53.55	37.14	69%	86.75% (0.26)	0.611 (n.s.)
	Late–partial		23	60.60	44.22	73%	83% (0.25)	
Between grades	Early–full	10	5	43.93	59.90	73%	84.4% (0.25)	0.754
	full	11	5	30.35	47.21	64%	89.1% (0.31)	(n.s.)
	Late–	10	12	47.33	61.11	77%	82.75% (0.17)	0.667
	partial	11	11	40.83	60.04	68%	83% (0.32)	(n.s.)

Note: n.s. = no statistically significant difference.

4.2.3 *Percentages of teacher and student talk*

The percentages of teacher[5] and student talk[6] in the observed lessons between school types and between grades are presented in Table 4.2.

A similar distribution of talk between teachers and their students was observed in the two types of EMI schools. For early–full EMI schools, teachers contributed 62% of the talk and students contributed only 8%. For late–partial EMI schools, teacher talk was 68% and student talk was only 5%. The remaining 27–30% of the talk was non-spoken interaction such as class activities.

Compared across grades, we see that in the early–full EMI schools, there was more teacher talk (1%) and less student talk (2%) in Grade 11. However, in the late–partial EMI schools, there was less teacher talk (9%) and less student talk (1%) in Grade 11.

The observed late–partial EMI schools appeared to be slightly more teacher-centred. Teacher talk contributed over 68% of the interaction time and student talk only contributed less than 5% of the interaction time.

For inferential statistics, the percentages of teacher talk and student talk between school types and between grades in each school group were compared using a Mann–Whitney U test in Table 4.3.

4.2.4 *Percentages of teacher talk*

For the comparison between school types, the percentages of teacher talk did not differ between the early–full EMI schools (md = 65.41%, $n = 10$) and the late–partial EMI schools (md = 68.86%, $n = 23$), $U = 105$, $z = -3.92$, $p = 0.695$, $r = 0.682$.[8]

Regarding the comparison between grades in each group, the percentages of teacher talk in the early–full EMI schools did not differ significantly between Grade 10 (md = 65.89%, $n = 5$) and Grade 11 (md = 64.93%, $n = 5$), $U = 11$, $z = -0.313$,

Table 4.2 Summary of teacher talk and student talk by school type and by grade

	School/ gradew		n (lessons)	Teacher talk		Student talk	
				Average interaction time per lesson (mins)	Percentages of interaction time[7]	Average interaction time per lesson (mins)	Percentages of interaction time
Between school types	Early– full		10	32.94	62%	4.20	8%
	Late– partial		23	41.36	68%	2.86	5%
Between grades	Early– full	G10	5	36.66	61%	7.27	12%
		G11	5	29.22	62%	1.12	2%
	Late–	G10	12	44.31	73%	3.02	5%
	partial	G11	11	38.15	64%	2.68	4%

Table 4.3 Mann–Whitney *U* test of the percentages of interaction time of teacher talk and student talk by school type and grade

	School/grade		*n* (lessons)	Teacher talk		Student talk	
				Median (SD)	*Asymp. Sig. (two-tailed)*	*Median (SD)*	*Asymp. Sig. (two-tailed)*
Between school types	Early– full		10	65.41% (0.25)	0.695 (n.s.)	9.18% (0.10)	0.147 (n.s.)
	Late– partial		23	68.86% (0.28)		4.63% (0.056)	
Between grades	Early– full	G10	5	65.89% (0.26)	0.754 (n.s.)	11.45% (0.11)	0.175 (n.s.)
		G11	5	64.93% (0.28)		6.91% (0.11)	
	Late– partial	G10	12	73.88% (0.18)	0.854	4.97% (0.050)	0.356
		G11	11	63.39% (0.36)	(n.s.)	4.26% (0.64)	(n.s.)

$p = 0.754$, $r = -0.099$. In the late–partial EMI schools, the percentage of teacher talk also did not differ significantly between Grade 10 (md = 73.88%, $n = 12$) and Grade 11 (md = 63.39%, $n = 11$), $U = 63$, $z = -0.185$, $p = 0.854$, $r = -0.039$.

4.2.5　*Percentages of student talk*

With respect to the comparison between school types, the percentages of student talk did not differ between the early–full EMI schools (md = 9.18%, $n = 10$) and the late–partial EMI schools (md = 4.63%, $n = 23$), $U = 78$, $z = -1.449$, $p = 0.147$, $r = 0.252$.

Concerning the comparison between grades in each group, the percentage of student talk in the early–full EMI schools did not differ significantly between Grade 10 (md = 11.45%, $n = 5$) and Grade 11 (md = 6.91%, $n = 5$), $U = 6$, $z = -1.358$, $p = 0.175$, $r = 0.429$. In the late–partial EMI schools, the percentage of student talk also did not differ significantly between Grade 10 (md = 4.97%, $n = 12$) and Grade 11 (md = 4.26%, $n = 11$), $U = 51$, $z = -0.923$, $p = 0.356$, $r = -0.192$.

Thus, the Mann–Whitney *U* test revealed that there is no statistically significant difference between early–full and late–partial EMI schools in the percentages of teacher and student talk. The same observation was made between Grades 10 and 11 in each school group.

4.2.6　*The number of initiations and responses*

Table 4.4 summarises the number and percentages of initiations and responses by teachers and students in the observed lessons between early–full and late–partial EMI schools and between Grades 10 and 11, reflecting the nature of interactive sequences in each type of school.

In the early–full EMI schools, there were a higher number of teacher initiations and responses but fewer numbers of student initiations and responses. This is different

Table 4.4 Mann–Whitney U test of the percentages of initiations and responses by teachers and students between Grades 10 and 11 in the two types of EMI schools

Between grades	Grade	Turns (% of interaction time)							
		Teachers				Students			
		Initiation	Asymp. Sig. (two-tailed)	Response	Asymp. Sig. (two-tailed)	Initiation	Asymp. Sig. (two-tailed)	Response	Asymp. Sig. (two-tailed)
Early–full	10	1271 (74%)	0.251 (n.s.)	240 (14%)	0.465 (n.s.)	52 (3%)	0.344 (n.s.)	158 (9%)	0.347 (n.s.)
	11	643 (74%)		129 (15%)		23 (3%)		78 (9%)	
Late–partial	10	2548 (67%)	0.902 (n.s.)	474 (12%)	0.518 (n.s.)	442 (12%)	0.265 (n.s.)	344 (9%)	0.712 (n.s.)
	11	2180 (70%)		311 (10%)		362 (12%)		240 (8%)	

Note: n.s. = no statistically significant difference.

from the late–partial EMI schools, with fewer teacher initiations and more student initiations and responses. From Grades 10 to 11, a similar proportion of initiations and responses between teachers and students were observed in the two types of schools.

Inferential statistics revealed that the percentages of teacher and students' initiations and responses in the early–full EMI schools did not differ significantly from those in the late–partial EMI schools. For the comparison between Grades 10 and 11 in each school group, the percentages of teacher and students' initiations and responses also show no significant difference.

4.2.7   Length of teacher and student turns

The total number of turns and the average length of teacher and student turns between the early–full and late–partial EMI schools are shown in Table 4.5. The table illustrates the total number of turns, the average length of each turn (in seconds), and the range of length for each group.

Both teachers and students in the late–partial EMI schools took more turns than those in the early–full EMI schools. However, the average length of teacher turns (42.03 s < 49.29 s) and student turns (3.16 s < 8.48 s) were slightly shorter. In addition, the maximum length of student turns in the late–partial EMI schools was four times shorter (302.6 s < 1,192.1 s) than teacher turns.

From Grades 10 to 11, the teacher and student average turn lengths were found to be different in the two types of schools. In the late–partial EMI schools, the average length of teacher turns was higher in Grade 11 than in Grade 10 (38.67 s versus 47.24 s) but the average length of student turns in Grade 11 was less than in Grade 10 (3.31 s versus 3.00 s). In the early–full EMI schools, the average length of teacher and student turns in Grade 11 was lower than in Grade 10 (61.44 s versus 39.49 s for teacher turns and 13.39 s versus 2.52 s for student turns). In

Table 4.5 Mann–Whitney *U* test of the percentages of initiations and responses by teachers and students between early–full and late–partial EMI schools

	School/ grade	n (lessons)	Teacher turns			Student turns		
			Total no. of turns per lesson	Average length of each turn per lesson (s)	Range of length of each turn per lesson (s)	Total no. of turns of per lesson	Average length of each turn per lesson (s)	Range of length of each turn per lesson (s)
Between school types	Early– full	10	40	49.29	0.1–1,215.1	30	8.48	0.1–1,192.1
	Late– partial	23	59	42.03	0.1–1,169.2	54	3.16	0.1–302.6
Between groups	Early– full	G10 5	36	61.44	0.2–1,192.1	33	13.39	0.1–410.3
		G11 5	44	39.49	0.1–1,215.1	27	2.52	0.1–48.1
	Late– partial	G10 12	69	38.67	0.1–1,169.2	54.8	3.31	0.1–302.6
		G11 11	48	47.24	0.1–823.5	53.6	3.00	0.1–58.5

Table 4.6 Mann–Whitney *U* test of the percentages of initiations and responses by teachers and students between Grades 10 and 11 in the two types of EMI schools

		n Lesson	Percentages of average length of teacher turns		Percentages of average length of student turns	
			Median (SD)	Asymp. Sig. (two-tailed)	Median (SD)	Asymp. Sig. (two-tailed)
Between school types	Early– full	10	2.45% (0.054)	0.176 (n.s.)	1.85% (0.13)	0.169 (n.s.)
	Late– partial	23	1.4% (0.022)		1% (0.011)	
Between grades	Early– full	G10 5	2.1% (0.058)	0.754 (n.s.)	4.3% (0.18)	0.116 (n.s.)
		G11 5	2.8% (0.055)		6% (0.021)	
	Late– partial	G10 12	1.65% (0.019)	0.579 (n.s.)	1% (0.014)	0.597 (n.s.)
		G11 11	1.4% (0.027)		0.9% (0.0047)	

Note: n.s. = no statistically significant difference.

addition, the maximum length of student turns in Grade 11 was less than in Grade 10 in both the early–full EMI schools (410.3 s versus 48.1 s) and the late–partial EMI schools (302.6 s versus 58.5 s).

Using inferential statistics, the percentages of the average length of teacher turns and of student turns between two types of schools and grades were compared using the Mann–Whitney *U* test, as shown in Table 4.6.

4.2.8   Teacher turns

For the comparison between school types, the percentages of the average length of teacher turns did not differ between the early–full EMI schools (md = 2.45%, *n* = 10) and the late–partial EMI schools (md = 1.4%, *n* = 23), $U = 80.5$, $z = -1.353$, $p = 0.176$, $r = 0.236$.

For the comparison between grades in each group, the percentages of mean length of teacher turn in the early–full EMI schools did not differ significantly between Grade 10 (md = 2.1%, $n = 5$) and Grade 11 (md = 2.8%, $n = 5$), $U = 11$, $z = -0.313$, $p = 0.754$, $r = 0.099$. In the late–partial EMI schools, the percentage of mean length of teacher turns also did not differ significantly between Grade 10 (md = 1.65%, $n = 12$) and Grade 11 (md = 1.4%, $n = 11$), $U = 57$, $z = -0.555$, $p = 0.579$, $r = 0.12$.

4.2.9   *Student turns*

Similarly, the percentages of mean length of student turns did not differ between the early–full EMI schools (md = 1.85%, $n = 10$) and the late–partial EMI schools (md = 1%, $n = 23$), $U = 80$, $z = -1.376$, $p = 0.169$, $r = 0.240$.

Additionally, the percentage of mean length of student turns in the early–full EMI schools did not differ significantly between Grade 10 (md = 4.3%, $n = 5$) and Grade 11 (md = 6%, $n = 5$), $U = 5.0$, $z = -1.571$, $p = 0.116$, $r = 0.50$. In the late–partial EMI schools, the percentage of mean length of student turns also did not differ significantly between Grade 10 (md = 1%, $n = 12$) and Grade 11 (md = 0.9%, n = 11), $U = 57.5$, $z = -0.528$, $p = 0.597$, $r = 0.11$.

Thus, the Mann–Whitney U test revealed that there is no statistically significant difference between early–full and late–partial EMI schools in the percentages of mean length of teacher turns and student turns. The same observation was made between Grades 10 and 11 in each school group. This corresponds to the descriptive statistics. In the early–full EMI schools, the average length of teacher turns in Grade 11 was much shorter than in Grade 10 (61.44 s versus 39.49 s).

4.2.10   *Types of teacher feedback*

A framework for analysing science teachers' feedback was adopted from Chin (2006). There are four types of teacher feedback in the follow-up or F-move of the IRF sequence identified in Chin's study. In her analysis of teacher–student interaction in junior (Grade 7) science classrooms in Singapore, Chin (2006) has further extended the 'F' in the IRF sequence model by classifying different types of feedback that teachers give to students' responses. According to Chin's view, teachers' feedback moves in the science classroom are distinctive moves in which teachers engage students in thinking about conceptual content. Feedback moves can therefore contribute to the construction of scientific knowledge.

A teacher's follow-up or feedback (F) in the IRF sequence can be in the form of a comment or statement followed by either another question or by further statements that expound more scientific content. Chin (2006) analysed these four major types of teacher feedback as follows:

1   affirmation-direct instruction
2   extension by responsive questioning: focusing and zooming
3   explicit correction, i.e., direct instruction
4   constructive challenge.

Table 4.7 Examples of science teachers' feedback

Feedback	Utterance	Examples
Affirmation-direct instruction	C-Q, C-S	S: Neutralisation T: Neutralisation, what else? (C-Q)
Extension by responsive questioning: focusing and zooming	C-Q	S: Water T: you think of water, does the rate of water have an effect? (C-Q)
Explicit correction, i.e., direct instruction	S-Q, S	S: Respiration. T: Respiration? No. Respiration is mitochondria, not ribosome
Constructive challenge	C-Q, Q	T: Why there is a negative acceleration? Yeah? What produces the negative acceleration?

Note: C, teacher comment; Q, teacher question; S, teacher statement.

Table 4.8 Percentages and the Mann–Whitney *U* test for the types of teacher feedback by school type

Between school types		Median (SD)		Mann–Whitney (Asymp. Sig. (two-tailed))
		Early–full	Late–partial	
Correct	Affirmation-direct instruction	2%	2%	0.937(n.s.)
Mixture of correct and incorrect	Extension by responsive questioning: focusing and zooming	2.5%	1%	0.410(n.s.)
Incorrect	Explicit correction-direct instruction	3.5%	1%	0.780(n.s.)
	Constructive challenge	< 1%	2%	0.102(n.s.)

Note: n.s. = no statistically significant difference.

Table 4.7 provides examples of these different types of science teachers' feedback from my data.

The percentages of science teacher feedback moves[9] in the observed lessons between early–full and late–partial EMI schools and between Grades 10 and 11 are shown in Tables 4.8 and 4.9.

A similar distribution of teachers' feedback was observed in both types of schools. Their frequency order was: 'affirmation-direct instruction' > 'extension by responsive questioning: focusing and zooming' > 'explicit correction' > 'constructive challenge'. From Grades 10 to 11, more 'affirmation-direct instruction' was observed in the early–full EMI schools but this pattern was reversed in the late–partial EMI schools.

With inferential statistics, there was no statistically significant difference between early–full and late–partial EMI schools regarding the percentages of teacher feedback. The same observation was made between Grades 10 and 11 in each group except in the late–partial EMI schools.

Table 4.9 Percentages and the Mann–Whitney *U* test on the types of teacher feedback by grade

| | | Median (SD) | | Mann–Whitney (Asymp. Sig. (two-tailed)) | Median (SD) | | Mann–Whitney (Asymp. Sig. (two-tailed)) |
| | | Early–full EMI | | | Late–partial EMI | | |
		G10	G11		G10	G11	
Correct	Affirmation-direct instruction	2% (0.03)	1% (0.05)	*p* = 0.819 (n.s.)	3% (0.03)	2% (0.02)	0.070 (n.s.)
Mixture of correct and incorrect	Extension by responsive questioning: focusing and zooming	4% (0.03)	1% (0.07)	*p* = 1.00 (n.s.)	2% (0.27)	1% (0.04)	0.704 (n.s.)
Incorrect	Explicit correction-direct instruction	5% (0.05)	2% (0.4)	*p* = 1.00 (n.s.)	1.5% (0.26)	1% (0.03)	0.527 (n.s.)
	Constructive challenge	< 1% (0.08)	< 1% (0.02)	*p* = 1.00 (n.s.)	2% (0.45)	2% (0.04)	0.417 (n.s.)

Note: n.s. = no statistically significant difference.

Table 4.10 Percentages and the Mann–Whitney *U* test on the wait time by school type and by grade

			n (lessons)	Average wait time per lesson (mins)	Percentages of wait time per lesson median (SD)	Mann–Whitney (Asymp. Sig. (two-tailed))
Between school types	Early–full		10	4.39	2.8% (0.19)	0.922 (n.s.)
	Late–partial		23	29.83	2.3% (0.05)	
Between grades	Early–full	G10	5	6.9	3.8% (0.26)	0.916 (n.s.)
		G11	5	1.87	1.8% (0.06)	
	Late–partial	G10	12	11.46	1.75% (0.04)	0.406 (n.s.)
		G11	11	19.05	3% (0.06)	

Note: n.s. = no statistically significant difference.

4.2.11   *Wait time*

The average and percentages of wait time[10] per lesson between teacher turns and student turns across schools and grades are presented in Table 4.10.

The percentage of wait time per lesson in the late–partial EMI schools (29.83 mins, 89%) was much longer than in the early–full EMI schools (4.39 mins, 11%). When comparing across the grades, in the early–full EMI schools, there was a lower average wait time (5.03 mins) in Grade 11 than in Grade 10. However, in late–partial EMI schools, there was a higher average wait time (7.59 mins) in Grade 11 than in Grade 10.

For inferential statistics, there is no statistically significant difference between early–full and late–partial EMI schools and between Grades 10 and 11 in percentages of wait time. Moreover, the percentage of wait time did not differ between the early–full EMI schools (md = 2.3%, n = 10) and the late–partial EMI schools (md = 2.8%, n = 23), U = 112.5, z = −0.098, p = 0.922, r = 0.02.

Similarly, the percentage of wait time in the early–full EMI schools did not differ significantly between Grade 10 (md = 3.8%, n = 5) and Grade 11 (gd = 1.8%, n = 5), U = 12, z = −0.105, p = 0.916, r = 0.03. In the late–partial EMI schools, the percentage of wait time also did not differ significantly between Grade 10 (gd = 1.75%, n = 12) and Grade 11 (md = 3%, n = 11), U = 52.5, z = −0.831, p = 0.406, r = 0.17.

4.2.12   Teacher and student talk in L1 and L2

Table 4.11 summarises the percentages of spoken L1 (Cantonese) and spoken L2 (English)[11] in the observed lessons between early–full and late–partial EMI schools and between Grades 10 and 11, reflecting the reality and extent of English medium instruction implemented in each school.

A high percentage of L2 in the interaction was observed in the early–full EMI schools, whereas a more equally distributed percentage of L1 and L2 spoken was observed in the late–partial EMI schools. There was more L1 use and less L2 use in Grade 11 than in Grade 10 in both types of schools.

Based on inferential statistics, the percentages of L1 and L2 spoken between school types and between grades in each school group were compared using a Mann–Whitney U test. For the comparison between school types, the percentages

Table 4.11 Percentages of L1 and L2 by school type and by grade

			Percentages of L2 spoken in the interaction		Percentages of L1 spoken in the interaction	
			Median (SD)	Mann– Whitney (Asymp. Sig. (two-tailed))	Median (SD)	Mann– Whitney (Asymp. Sig. (two-tailed))
Between school types	Early– full		93.5% (0.13)	$p < 0.0001$****	6.5% (0.13)	$p < 0.0001$****
	Late– partial		44% (0.25)		53% (0.25)	
Between grades	Early– full	G10	< 100% (0.18)	0.665 (n.s.)	< 1% (0.18)	0.665 (n.s.)
		G11	91% (0.07)		9% (0.07)	
	Late– partial	G10	47.5% (0.20)	0.268 (n.s.)	52.5% (0.20)	0.712 (n.s.)
		G11	44% (0.29)		56% (0.30)	

Note: ****$p < 0.0001$; n.s. = no statistically significant difference.

of L2 spoken differed between the early–full EMI schools (md = 93.5%, n = 10) and the late–partial EMI schools (md = 44%, n = 23), U = 11, z = −4.079, p < 0.0001, r = 0.71.[12] This would be considered as a medium effect size using Cohen's (1988) criterion (medium effect = 0.5).

The same observation was found for the percentage of L1 spoken between the early–full EMI schools (md = 6.5%, n = 10) and the late–partial EMI schools (md = 53%, n = 23), U = 19, z = −3.768, p <.0001, r = 0.66. This is also considered a medium effect size using Cohen's (1988) criterion.

For the comparison between Grades 10 and 11 in each school group, the percentage of L1 spoken and L2 spoken shows no significant difference. Specifically, the percentage of L2 spoken in the early–full EMI schools did not differ significantly between Grade 10 (md = < 100%, n = 5) and Grade 11 (md = 91%, n = 5), U = 10.5, z = −0.432, p = 0.665, r = −0.14. In the late–partial EMI schools, the percentage of L2 spoken also did not differ significantly between Grade 10 (md = 47.5%, n = 12) and Grade 11 (md = 44%, n = 11), U = 48, z = −1.109, p = 0.268, r = −0.23.

Furthermore, the percentage of L1 spoken in the early–full EMI schools did not differ significantly between Grade 10 (md = <1%, n = 5) and Grade 11 (md = 9%, n = 5), U = 10.5, z = −0.432, p = 0.665, r = −0.14. This was also seen in the late–partial EMI schools; the percentage of L1 spoken also did not differ significantly between Grade 10 (md = 52.5%, n = 12) and Grade 11 (md = 56%, n = 11), U = 60, z = −0.370, p = 0.712, r = −0.078.

Table 4.12 summarises the percentages of L1 and L2 use by teachers and students between early–full and late–partial EMI schools and between Grades 10 and 11.

4.2.13   Teachers' L1 and L2 use

Based on Table 4.12, a high percentage of L2 use was observed in the early–full EMI schools, whereas a high percentage of L1 use was observed in the late–partial EMI schools.

For inferential statistics, the percentages of L1 use by teachers differed relatively significantly between the early–full EMI schools (md = 1.5%, n = 10) and the late–partial EMI schools (md = 26%, n = 23), U = 29.5, z = −3.361, p = 0.001. r = 0.59, making it a medium effect size using Cohen's (1988) criterion.

Similarly, the percentages of L2 use by teachers differed between the early–full EMI schools (md = 54%, n = 10) and the late–partial EMI schools (md = 42%, n = 23), U = 47.5, z = −2.647, p = 0.008, r = 0.46. This is considered a small effect size based on Cohen's (1988) criterion (small effect = 0.2).

When comparing across grades, the percentage of L1 use by teachers in the early–full EMI schools did not differ significantly between Grade 10 (md = <1%, n = 5) and Grade 11 (md = 3%, n = 5), U = 11, z = −0.334, p = 0.738, r = −0.11. In the late–partial EMI schools, the percentage of L1 use by teachers also did not differ significantly between Grade 10 (md = 42.5%, n = 12) and Grade 11 (md = 21%, n = 11), U = 52, z = −0.863, p = 0.388, r = −0.18.

Table 4.12 Percentage of L1 and L2 use by teachers and students by school type and by grade

Between school types	Teachers			Students		
	Median (SD)		*Mann–Whitney (Asymp. Sig. (two-tailed))*	*Median (SD)*		*Mann–Whitney (Asymp. Sig. (two-tailed))*
	Early–full	*Late–partial*		*Early–full*	*Late–partial*	
L1 use	1.5% (.09)	26% (0.20)	0.001**	1% (0.06)	8% (0.22)	0.079*
L2 use	54% (.09)	42% (0.19)	0.008*	33% (0.10)	7% (0.11)	p < 0.0001****

Between grades	Teachers				Students			
	Early–full		*Late–partial*		*Early–full*		*Late–partial*	
	Median (SD)	*Mann–Whitney (Asymp. Sig. (two-tailed))*	*Median (SD)*	*Mann–Whitney (Asymp. Sig. (two-tailed))*	*Median (SD)*	*Mann–Whitney (Asymp. Sig. (two-tailed))*	*Median (SD)*	*Mann–Whitney (Asymp. Sig. (two-tailed))*
L1 G10	<1% (.12)	0.738 (n.s.)	42.5% (0.18)	0.388 (n.s.)	<1% (0.06)	0.448 (n.s.)	6.5% (0.15)	0.440 (n.s.)
G11	3% (0.04)		21% (0.22)		1% (0.06)		12% (0.28)	
L2 G10	52% (0.10)	0.834 (n.s.)	41% (0.14)	0.601 (n.s.)	33% (0.13)	0.753 (n.s.)	7.5% (0.11)	0.665 (n.s.)
G11	56% (0.09)		42% (0.24)		33% (0.07)		6% (0.11)	

Note: *p < 0.05; **p < 0.01; ***p < 0.001; ****p < 0.0001; n.s. = no statistically significant difference.

Furthermore, the percentage of L2 use by teachers in the early–full EMI schools did not differ significantly between Grade 10 (md = 52%, n = 5) and Grade 11 (md = 56%, n = 5), U = 11.5, z = −21, p = 0.834, r = −0.04. The percentage of L2 use by teachers in the late–partial EMI schools also did not differ significantly between Grade 10 (md = 41%, n = 12) and Grade 11 (md = 42%, n = 11), U = 57.5, z = −0.524, p = 0.601, r = −0.11.

4.2.14   *Students' L1 and L2 use*

A relatively high percentage of L2 use was observed in the early–full EMI schools. A similar percentage of L1 and L2 spoken was observed in the late–partial EMI schools. From Grades 10 to 11, there was a similar ratio of L1 and L2 use in both types of schools.

For inferential statistics, the percentages of L1 use by students differed between the early–full EMI (md = 1%, n = 10) and the late–partial EMI schools (md = 8%, n = 23), U = 70.5, z = −1.754, p = 0.079, r = −0.31. This is considered a small effect size using Cohen's (1988) criterion.

Similarly, the percentages of L2 use by students differed between the early–full EMI schools (md = 33%, n = 10) and the late–partial EMI schools (md = 7%, n = 23), U = 13, z = −4.002, p < 0.0001, r = −0.70, making it a medium effect size based on Cohen's (1988) criterion. Thus, the L1 and L2 use by students illustrates a statistically significant difference between early–full and late–partial EMI schools.

When comparing across grades, the percentage of L1 use by students in the early–full EMI schools did not differ significantly between Grade 10 (md = < 1%, n = 5) and Grade 11 (md = 1%, n = 5), U = 9, z = −0.759, p = 0.448, r = −0.24. In the late–partial EMI schools, the percentage of L1 use by students also did not differ significantly between Grade 10 (md = 33%, n = 12) and Grade 11 (md = 33%, n = 11), U = 53.5, z = −0.772, p = 0.440, r = −0.16. The percentage of L2 use by students in the early–full EMI schools did not differ significantly between Grade 10 (md = 33%, n = 5) and Grade 11 (md = 33%, n = 5), U = 11, z = −0.315, p = 0.753, r = −0.01. In the late–partial EMI schools, the percentage of L2 use by students also did not differ significantly between Grade 10 (md = 7.5%, n = 12) and Grade 11 (md = 6%, n = 11), U = 59, z = −0.432, p = 0.665, r = −0.09.

Table 4.13 summarises the percentages of inter- and intra-sentential code-switching for both teachers and students between early–full and late–partial EMI schools and between Grades 10 and 11.

Each percentage refers to the number of occurrences in each speaker's turn. For code-switching patterns, the intra-sentential pattern means code-switching occurs within a sentence, thereby mixing two languages in a sentence. For example, during a chemistry lesson in a late–partial EMI school, the teacher said, '*Our focus today is* 中和作用 (*translation: neutralisation*) *that we briefly discussed yesterday*'. Alternatively, the inter-sentential pattern refers to code-switches that happen in between sentences, mixing L1 and L2 sentences. For example, the same teacher explained, '*Neutralisation is a chemical reaction*

Table 4.13 Percentages of inter- and intra-sentential code-switching by school type and by grade

			Intra-sentential (code-switching within a sentence)		Inter-sentential (code-switching between sentences)	
			Median (SD)	*Mann–Whitney (Asymp. Sig. (two-tailed))*	*Median (SD)*	*Mann–Whitney (Asymp. Sig. (two-tailed))*
Between school types	Early–full		< 1% (0.22)	0.002**	23% (0.38)	0.254 (n.s.)
	Late–partial		48% (0.23)		47% (0.24)	
Between grades	Early–full	G10	< 1% (0.20)	0.814 (n.s.)	< 1% (0.37)	0.577 (n.s.)
		G11	< 1% (0.26)		46% (0.43)	
	Late–partial	G10	54% (0.17)	0.185 (n.s.)	46% (0.17)	0.579 (n.s.)
		G11	39% (0.27)		53% (0.30)	

Note: **$p < 0.01$; n.s. = no statistically significant difference.

in which an acid and a base react quantitatively with each other to form salts and water.' She then continued with the same expression in Cantonese and said, '中和作用是指酸和鹼反應生成鹽和水的反應。' (translation: *Neutralisation refers to the reaction between acid and base, which reacts to form salts and water*).

When comparing schools, a statistical difference was found for the intra-sentential code-switching between early–full (md = <1%, n = 10) and late–partial EMI schools (md = 48%, n = 23), U = 38, z = –3.032, p = 0.002, r = –0.52. This is considered as a medium effect size according to Cohen's (1988) criterion. For other comparison groups, no statistical difference was found for either intra-sentential or inter-sentential code-switching between school types and between Grades 10 and 11 in each group.

However, the descriptive statistics show that teachers and students in early–full EMI schools tended to have more inter-sentential code-switching (in between sentences). Those in late–partial EMI schools tended to have more intra-sentential code-switching (within a sentence), although the difference here is not large when compared with the early–full EMI schools. From Grades 10 to 11 across both types of schools, teachers and students tended to use more inter-sentential code-switching relative to intra-sentential code-switching.

4.2.15   *The pedagogical functions of teacher and student talk*

The analytical framework used here was adopted from Tsui's (1985) 17-category system with the additional seven categories added by Lo and Macaro (2012). These categories were explained in Chapter 3. In this structured coding scheme, the pedagogical functions of teacher and student talk were coded and compared between school types and grades. Table 4.14 provides examples of the six most frequent pedagogical functions of acts, based on the 17-category system based on Tsui's definitions, found in the observed lessons.

The frequency, percentages, and proportion of interaction time in both school types are summarised in Table 4.15. Tables 4.16 and 4.17 summarise the most

Table 4.14 Examples of different types of science teachers' feedback

Pedagogical functions	Examples
Inform	T: *That is MRSA, very famous bacteria*
Self-utterance	T: *All are correct? Really? No! (the teacher talked to himself)*
Direct	T: *Read the question carefully*
Reply (Student)	T: *The amount of antibiotic A B C D are they different?*
	S: *B can kill more bacteria*
Check	T: *Okay? Any questions about this petri dish? Any questions?*
Express Opinions	T: *Sir, it is too much water*

Table 4.15 Most frequently occurring pedagogical functions of talk by school type

Between school types	Early–full EMI schools		Late–partial EMI schools		Mann–Whitney (Asymp. Sig. (two-tailed))	
	Frequency (n)	Median (SD)	Frequency (n)	Median (SD)	Original p value	Adjusted p value (with alpha =.008)
Inform	772	2.5% (0.02)	1,938	3.4% (0.01)	0.938 (n.s.)	0.156 (n.s.)
Self-utterance	274	3.3% (0.02)	686	2.7% (0.03)	0.543 (n.s.)	0.091 (n.s.)
Direct	239	2.8% (0.01)	683	3% (0.02)	0.739 (n.s.)	0.123 (n.s.)
Reply	236	4.7% (0.0001)	583	4.7% (0.0001)	p = 1.00 (n.s.)	0.167 (n.s.)
Check	224	2.4% (0.05)	305	1.9% (0.02)	0.583 (n.s.)	0.097 (n.s.)
Express Opinions	43	0.5% (0.01)	535	0.5% (0.10)	0.608 (n.s.)	0.101 (n.s.)

Note: n.s. = no statistically significant difference.

Table 4.16 Most frequently occurring pedagogical functions of talk by grade in early–full EMI schools

Between grades	Early–full EMI school				Mann–Whitney (Asymp. Sig. (two-tailed))	
	Grade 10		Original p value		Original p value	Adjusted p value (with alpha =.008)
	Frequency (n)	Median (SD)	Frequency (n)	Median (SD)		
Inform	497	2.8% (0.02)	275	1.7% (0.02)	0.754 (n.s.)	0.126 (n.s.)
Self-utterance	193	3.4% (0.02)	81	1.4% (0.02)	0.53 (n.s.)	0.088 (n.s.)
Direct	166	3.1% (0.01)	73	2.5% (0.01)	0.675 (n.s.)	0.113 (n.s.)
Reply	158	4.7% (0.0001)	78	4.7% (0.06)	p = 1.00 (n.s.)	0.167 (n.s.)
Check	129	2.7% (0.05)	95	2.1% (0.06)	0.917 (n.s.)	0.153 (n.s.)
Express Opinions	31	1% (0.01)	12	0.3% (0.01)	0.289 (n.s.)	0.048 (n.s.)

Note: n.s. = no statistically significant difference.

Table 4.17 Most frequently occurring pedagogical functions of talk by grade in late–partial EMI schools

Between grades	Late–partial EMI schools				Mann–Whitney (Asymp. Sig. (two-tailed))	
	Grade 10		Grade 11		Original p value	Adjusted p value (with alpha = 0.008)
	Frequency (n)	Median (SD)	Frequency (n)	Median (SD)		
Inform	1036	3.9% (0.01)	902	2.3% (0.02)	0.441 (n.s.)	0.0735 (n.s.)
Self-utterance	281	2.8% (0.02)	405	1.1% (0.04)	0.295 (n.s.)	0.0492 (n.s.)
Direct	410	3.4% (0.02)	273	3% (0.02)	0.388 (n.s.)	0.0647 (n.s.)
Reply	344	4.7% (0.0001)	239	4.7% (0.0001)	p = 1.00 (n.s.)	0.167 (n.s.)
Check	181	3.3% (0.02)	124	1.9% (0.02)	0.441 (n.s.)	0.0735 (n.s.)
Express Opinions	280	0.8% (0.09)	255	0.5% (0.10)	0.757 (n.s.)	0.126 (n.s.)

Note: n.s. = no statistically significant difference.

frequently occurring pedagogical functions of talk by grade in early–full and late–partial EMI schools, respectively.

The Mann–Whitney *U* test was utilised for between-group comparisons with Bonferroni corrections for multiple comparisons ($n = 6$). The corrected alpha value is 0.008, where each of the *p* values was divided by 6; the corrected *p* values were listed in the table. No statistical difference was found for the five frequent pedagogical functions between early–full and late–partial EMI schools. However, from the descriptive statistics, the four most frequent pedagogical functions in teacher and student talk were the same in both types of schools: inform > self-utterance > direct > reply. The last three pedagogical functions were all fairly close in percentages. However, the fifth most frequent function in the early–full EMI schools was 'check', whereas 'express opinion' was the fifth most frequent function in the late–partial EMI schools. From Grades 10 to 11, a similar pattern of ranking was observed. In the early–full EMI schools, 'self-utterance' was more frequent than 'check' in Grade 10 but this pattern was reversed in Grade 11. In the late–partial EMI schools, 'direct' was more common than 'self-utterance', but this was again reversed in Grade 11.

4.2.16 *Teachers' questions*

Adopting Yip's (2004) taxonomy of teacher questions as well as the framework for analysis of said questions by Lo and Macaro (2012), it was revealed that the majority of teacher questions across both school types were factual-type questions. However, upon further analysis, the questions asked by science teachers differed in terms of order according to Lo and Macaro's (2012) framework.

4.2.16.1 *Teachers' questions in content-subject classrooms*

The framework for analysing teachers' questions in content-subject classrooms was adopted from Lo and Macaro (2012), as explained in Chapter 3. Table 4.18

illustrates some examples of different teachers' questions in content-subject class-rooms. Tables 4.19 and 4.20 summarise the types of teachers' questions found in the observed lessons between the early–full and late–partial EMI schools and be-tween Grades 10 and 11. The Mann–Whitney U test for between-group compari-sons with Bonferroni corrections for multiple comparisons ($n = 8$) was conducted. The corrected alpha value is 0.006; each of the p values was divided by 8 and the corrected p values were listed in the table. No statistical difference was found for the five frequent pedagogical functions between early–full and late–partial EMI schools and between Grades 10 and 11 in each group.

The only statistical difference is found for the information type of genuine question between early–full (md = 4.6%, $n = 10$) and late–partial EMI schools

Table 4.18 Examples of teachers' questions in content-subject classrooms

Types of questions	Examples
Factual questions	What else do you know about antibiotics?
Yes/no questions	Do we need to rinse the conical flask with sodium hydroxide?
Alternative questions	Net uptake or net release?
Reasoning questions	Why does the mitotic cell division have the same genetic material?
Explaining word meaning questions	What is the meaning of saturated?
Opining questions	What do you think about the consequence of combustion?
Information questions	Have you drawn the result?
Restating elicits	Can we find something like this? Where? Any example of the plant without chlorophyll?

Table 4.19 Types of teachers' questions by school and by grade

Types of questions		Between school types			
		Median (SD)		Mann–Whitney (Asymp. Sig. (two-tailed))	
Teachers' elicits		Early–full	Late–partial	Original	Adjusted p value (with alpha = 0.006)
Display questions	Factual questions	1.7% (0.04)	2.2% (0.03)	0.953 (n.s.)	0.119 (n.s.)
	Yes/no questions	0.6% (0.02)	2.2% (0.05)	0.288 (n.s.)	0.036 (n.s.)
	Alternative questions	1.9% (0.04)	1.3% (0.04)	0.763 (n.s.)	0.0954 (n.s.)
	Reasoning	1.9% (0.06)	0.9% (0.03)	0.259 (n.s.)	0.0324 (n.s.)
	Explaining word meanings	< 1% (0.04)	< 1% (0.07)	0.499 (n.s.)	0.0624 (n.s.)
Genuine questions	Opinions	< 1% (0.0001)	< 1% (0.0001)	p = 1.00 (n.s.)	0.0125 (n.s.)
	Information	4.6% (0.05)	< 1% (0.03)	0.033	0.00413*
Restating elicits		1.2% (0.04)	1.6% (0.03)	0.937 (n.s.)	0.0117

Note: *$p < 0.00625$; n.s. = no statistically significant difference.

Quantitative analysis of classroom interactions 105

Table 4.20 Teachers' questions compared by school type and grades

Types of questions		Between grades							
		Early-full				Late-partial			
Teachers' elicits		Median (SD)		Mann–Whitney (Asymp. Sig. (two-tailed))		Median (SD)		Mann–Whitney (Asymp. Sig. (two-tailed))	
		G10	G11	Original	Adjusted p value (with alpha =.006)	G10	G11	Original	Adjusted p value (with alpha = 0.006)
Display questions	Factual questions	1.9% (0.03)	1.4% (0.05)	0.599 (n.s.)	0.749 (n.s.)	2.7% (0.02)	1.2% (0.03)	0.218 (n.s.)	0.0273 (n.s.)
	Yes/no questions	<1% (0.02)	1.1% (0.02)	0.432 (n.s.)	0.054 (n.s.)	2.2% (0.06)	1.1% (0.03)	0.511 (n.s.)	0.0639 (n.s.)
	Alternative questions	1.3% (0.04)	2.5% (0.05)	0.831 (n.s.)	0.104 (n.s.)	2.5% (0.04)	1.3% (0.03)	0.145 (n.s.)	0.0181 (n.s.)
	Reasoning	1.9% (0.05)	1.9% (0.07)	0.832 (n.s.)	0.104 (n.s.)	0.9% (0.04)	0.9% (0.02)	0.655 (n.s.)	0.0819 (n.s.)
	Explaining word meanings	<1% (0.04)	<1% (0.04)	p = 1.00 (n.s.)	0.125 (n.s.)	<1% (0.08)	<1% (0.06)	0.873 (n.s.)	0.109 (n.s.)
Genuine questions	Opinions	< 1% (0.0001)	< 1% (0.0001)	p = 1.00 (n.s.)	0.125 (n.s.)	< 1% (0.0001)	< 1% (0.0001)	p = 1.00 (n.s.)	0.125 (n.s.)
	Information	1.8% (0.06)	7.3% (0.04)	0.831 (n.s.)	0.104 (n.s.)	0.9% (0.04)	<1% (0.03)	0.453 (n.s.)	0.0566 (n.s.)
Restating elicits		1.9% (0.04)	0.3% (0.06)	0.344 (n.s.)	0.043 (n.s.)	1.8% (0.03)	1.6% (0.03)	0.829 (n.s.)	0.0104 (n.s.)

Note: n.s. = no statistically significant difference.

(md = 1%, n = 23), U = 63.5, z = −2.132, p < 0.033, r = −0.37. This is considered a small effect size using Cohen's (1988) criterion.

The descriptive statistics show that most of the teachers' questions were 'factual' (over 44%), followed by 'restating elicits' (over 27%) in the two types of schools. Among all the sub-types in display questions, more 'reasoning' questions (11%) were observed in the early–full EMI schools. More 'yes/no' questions (9%) were observed in the late–partial EMI schools.

From Grades 10 to 11, the distributions of 'factual' questions and 'restating elicits' in the early–full EMI schools were similar. However, in the late–partial EMI schools, the number of factual questions in Grade 11 (43%) was less than in Grade 10 (46%) and the 'restating elicits' in Grade 11 (26%) was also less than in Grade 10 (31%). One noticeable change in the two groups was that in the early–full EMI schools, there were higher percentages of the three types of questions in Grade 11 than in Grade 10: 'yes/no' questions, 'alternative' questions and 'information' questions. The opposite was the case in the late–partial EMI schools, as teachers from both types of schools used more 'reasoning' questions.

4.2.16.2   *Science teachers' questions*

The framework for analysing science teachers' questions in Hong Kong classrooms was adopted from Yip (2004), which Chapter 3 goes into more detail about. Table 4.21 provides examples of science teachers' questions. Tables 4.22 and 4.23 summarise the types of teachers' questions between early–full and late–partial EMI schools and between Grades 10 and 11. Mann–Whitney U test for between-groups comparisons with Bonferroni correction for multiple comparisons (n = 14). The corrected alpha value is 0.004. Each of the p values was divided by 14, and the corrected p values were listed in the table (Tables 4.24 and 4.25).

Table 4.21 Examples of science teachers' questions

Types of science teachers' questions	Examples
Recalling	*What do you know about antibiotics? When you learned liberal studies?*
Explanation	*Would you explain what happened? At that time, what happened to the basketball at t = 0?*
Analysis	*In this experiment, which part is the experimental setup?*
Evaluation	*Do you think cholesterol is good or not, to be honest?*
Synthesis	*So if this compound is only slightly soluble in water, how can we make a solution that is clear?*
Motivation	*So you can, what can you do for the protection of the environment?*
Eliciting	*When the bacteria inside the place seems bad, do you know what the benefits are?*
Challenging	*Can you find some plant without chlorophyll? Can you find some plants that the whole plant is not green in colour?*
Extending	*How do you know how much chlorophyll is needed in photosynthesis? Do we need a comparison?*
Application	*What are the chemicals inside the glass cleaner, or window cleaner?*

Table 4.22 Types of science teachers' questions by school type and by grade

Teachers' science questions		Between school types			
		Median (SD)		Mann–Whitney (Asymp. Sig. (two-tailed))	
		Early–full	Late–partial	Original	Adjusted p value (with alpha =.004)
Lower order	Lower order	3% (0.05)	2% (0.02)	0.749 (n.s.)	0.0535 (n.s.)
	Higher order	2% (0.02)	3% (0.03)	0.520 (n.s.)	0.0371 (n.s.)
	Motivation	3.5% (0.04)	2% (0.03)	0.576 (n.s.)	0.0411 (n.s.)
	Conceptual change	1.5% (0.05)	2% (0.03)	0.910 (n.s.)	0.0650 (n.s.)
Lower order	Recalling	3% (0.05)	2% (0.02)	0.749 (n.s.)	0.0535 (n.s.)
	Explanation	< 1% (0.05)	< 1% (0.04)	0.996 (n.s.)	0.0711 (n.s.)
Higher order	Analysis	1% (0.02)	2% (0.06)	0.897 (n.s.)	0.0641 (n.s.)
	Evaluation	1.5% (0.02)	3% (0.03)	0.271 (n.s.)	0.0194 (n.s.)
	Synthesis	2% (0.03)	2% (0.03)	$p = 1.00$ (n.s.)	0.0714 (n.s.)
Motivation	Motivation	3.5% (0.04)	2% (0.03)	0.576 (n.s.)	0.0411 (n.s.)
Conceptual change	Eliciting	< 1% (0.02)	< 1% (0.09)	0.984 (n.s.)	0.0703 (n.s.)
	Challenging	< 1% (0.04)	< 1% (0.06)	$p = 1.00$ (n.s.)	0.0714 (n.s.)
	Extending	2.5% (0.06)	1% (0.04)	0.338 (n.s.)	0.0241 (n.s.)
	Application	2% (0.06)	< 1% (0.04)	0.520 (n.s.)	0.0371 (n.s.)

Note: n.s. = No statistically significant difference.

Table 4.23 Types of science teachers' questions by school type

Teachers' science questions		Early–full			Late–partial		
		Median (SD)		Mann–Whitney (Asymp. Sig. (two-tailed))	Median (SD)		Mann–Whitney (Asymp. Sig. (two-tailed))
		G10	G11	Adjusted p value (with alpha =.004)	G10	G11	Adjusted p value (with alpha =.004)
Lower order	Lower order	3% (0.03)	< 1% (0.07)	0.0324 (n.s.)	3% (0.03)	2% (0.01)	0.0084 (n.s.)
	Higher order	2% (0.02)	2% (0.01)	0.0536 (n.s.)	3.5% (0.03)	3% (0.03)	0.0239 (n.s.)
	Motivation	4% (0.06)	3% (0.03)	0.0481 (n.s.)	2% (0.03)	1% (0.03)	0.0555 (n.s.)
	Conceptual change	2% (0.04)	1% (0.06)	0.0654 (n.s.)	1.5% (0.04)	2% (0.03)	0.0554 (n.s.)
Lower order	Recalling	3% (0.03)	< 1% (0.0001)	0.0327 (n.s.)	3.5% (0.03)	2% (0.01)	0.0069 (n.s.)
	Explanation	< 1% (0.05)	<1% (0.0001)	0.0581 (n.s.)	< 1% (0.05)	1% (0.04)	0.0392 (n.s.)
Higher order	Analysis	1% (0.01)	1% (0.02)	0.0654 (n.s.)	1.5% (0.06)	3% (0.06)	0.0696 (n.s.)
	Evaluation	2% (0.03)	<1% (0.02)	0.0273 (n.s.)	4.5% (0.03)	2% (0.03)	0.0217 (n.s.)
	Synthesis	2% (0.03)	2% (0.03)	0.0654 (n.s.)	3.5% (0.03)	1% (0.04)	0.0096 (n.s.)
Motivation	Motivation	4% (0.06)	3% (0.03)	0.0481 (n.s.)	2% (0.03)	1% (0.03)	0.0555 (n.s.)

(Continued)

Table 4.23 (Continued)

Teachers' science questions		Early–full		Mann–Whitney (Asymp. Sig. (two-tailed))	Late–partial		Mann–Whitney (Asymp. Sig. (two-tailed))
		Median (SD)			Median (SD)		
		G10	G11	Adjusted p value (with alpha =.004)	G10	G11	Adjusted p value (with alpha =.004)
Conceptual change	Eliciting	< 1% (0.02)	< 1% (0.03)	0.0366 (n.s.)	< 1% (0.03)	< 1% (0.13)	0.0299 (n.s.)
	Challenging	< 1% (.001)	< 1% (0.06)	0.0097 (n.s.)	< 1% (0.07)	< 1% (0.04)	0.0226 (n.s.)
	Extending	5% (0.05)	< 1% (0.07)	0.0651 (n.s.)	0.5% (0.05)	1% (0.02)	0.0658 (n.s.)
	Application	2% (0.06)	4% (0.06)	0.0279 (n.s.)	< 1% (0.05)	< 1% (0.03)	0.0557 (n.s.)

Note: n.s. = no statistically significant difference.

Table 4.24 Types of science teachers' questions by school type and by grade

Teachers' science questions		Between school types		Mann–Whitney (Asymp. Sig. (two-tailed))	
		Median (SD)			
		Early–full	Late–partial	Original	Adjusted p value (with alpha = 0.004)
	Lower order	3% (0.05)	2% (0.02)	0.749 (n.s.)	0.0535 (n.s.)
	Higher order	2% (0.02)	3% (0.03)	0.520 (n.s.)	0.0371 (n.s.)
	Motivation	3.5% (0.04)	2% (0.03)	0.576 (n.s.)	0.0411 (n.s.)
	Conceptual change	1.5% (0.05)	2% (0.03)	0.910 (n.s.)	0.0650 (n.s.)
Lower order	Recalling	3% (0.05)	2% (0.02)	0.749 (n.s.)	0.0535 (n.s.)
	Explanation	< 1% (0.05)	< 1% (0.04)	0.996 (n.s.)	0.0711 (n.s.)
Higher order	Analysis	1% (0.02)	2% (0.06)	0.897 (n.s.)	0.0641 (n.s.)
	Evaluation	1.5% (0.02)	3% (0.03)	0.271 (n.s.)	0.0194 (n.s.)
	Synthesis	2% (0.03)	2% (0.03)	p = 1.00 (n.s.)	0.0714 (n.s.)
Motivation	Motivation	3.5% (0.04)	2% (0.03)	0.576 (n.s.)	0.0411 (n.s.)
Conceptual change	Eliciting	< 1% (0.02)	< 1% (0.09)	0.984 (n.s.)	0.0703 (n.s.)
	Challenging	< 1% (0.04)	< 1% (0.06)	p = 1.00 (n.s.)	0.0714 (n.s.)
	Extending	2.5% (0.06)	1% (0.04)	0.338 (n.s.)	0.0241 (n.s.)
	Application	2% (0.06)	< 1% (0.04)	0.520 (n.s.)	0.0371 (n.s.)

Note: n.s. = no statistically significant difference.

No statistical difference was found for the science teachers' questions between early–full and late–partial EMI schools, and between Grades 10 and 11 in each school group. However, for the descriptive statistics, more recalling questions were observed in the early–full EMI schools than in the late–partial EMI schools.

Table 4.25 Types of science teachers' questions by school type

Teachers' science questions		Early–full			Late–partial		
		Median (SD)		Mann–Whitney (Asymp. Sig. (two-tailed))	Median (SD)		Mann–Whitney (Asymp. Sig. (two-tailed))
		G10	G11	Adjusted p value (with alpha =.004)	G10	G11	Adjusted p value (with alpha =.004)
	Lower order	3% (0.03)	< 1% (0.07)	0.0324 (n.s.)	3% (0.03)	2% (0.01)	0.0084 (n.s.)
	Higher order	2% (0.02)	2% (0.01)	0.0536 (n.s.)	3.5% (0.03)	3% (0.03)	0.0239 (n.s.)
	Motivation	4% (0.06)	3% (0.03)	0.0481 (n.s.)	2% (0.03)	1% (0.03)	0.0555 (n.s.)
	Conceptual change	2% (0.04)	1% (0.06)	0.0654 (n.s.)	1.5% (0.04)	2% (0.03)	0.0554 (n.s.)
Lower order	Recalling	3% (0.03)	< 1% (0.0001)	0.0327 (n.s.)	3.5% (0.03)	2% (0.01)	0.0069 (n.s.)
	Explanation	< 1% (0.05)	<1% (0.0001)	0.0581 (n.s.)	< 1% (0.05)	1% (0.04)	0.0392 (n.s.)
Higher order	Analysis	1% (0.01)	1% (0.02)	0.0654 (n.s.)	1.5% (0.06)	3% (0.06)	0.0696 (n.s.)
	Evaluation	2% (0.03)	<1% (0.02)	0.0273 (n.s.)	4.5% (0.03)	2% (0.03)	0.0217 (n.s.)
	Synthesis	2% (0.03)	2% (0.03)	0.0654 (n.s.)	3.5% (0.03)	1% (0.04)	0.0096 (n.s.)
Motivation	Motivation	4% (0.06)	3% (0.03)	0.0481 (n.s.)	2% (0.03)	1% (0.03)	0.0555 (n.s.)
Conceptual change	Eliciting	< 1% (0.02)	< 1% (0.03)	0.0366 (n.s.)	< 1% (0.03)	< 1% (0.13)	0.0299 (n.s.)
	Challenging	< 1% (0.001)	< 1% (0.06)	0.0097 (n.s.)	< 1% (0.07)	< 1% (0.04)	0.0226 (n.s.)
	Extending	5% (0.05)	< 1% (0.07)	0.0651 (n.s.)	0.5% (0.05)	1% (0.02)	0.0658 (n.s.)
	Application	2% (0.06)	4% (0.06)	0.0279 (n.s.)	< 1% (0.05)	< 1% (0.03)	0.0557 (n.s.)

Note: n.s. = no statistically significant difference.

In the late–partial EMI schools, there was more frequent use of higher-order questions (analysis, evaluation) and motivation questions than in the early–full EMI schools. There were more lower-order and higher-order questions in the early–full EMI schools in Grade 11 than in Grade 10. Yet, there were more higher-order questions in the late–partial EMI schools across Grades 10 and 11.

4.3 Summary of the quantitative findings

The quantitative comparison provides a representative sample of classroom interactions in Hong Kong's EMI secondary schools using a number of objective, comprehensive, and systematic methods. The major findings were categorised as follows: interaction time, distribution of teacher and student talk, the average length of teacher and student turns, wait time between teacher and student talk, the ratio of L1 and L2 use between teachers and students, code-switching patterns, the pedagogical function of teachers' and student' turns, and the types of teachers' questions.

4.3.1   Interaction time, teacher and student talk, average length of teacher and student turns

There was no statistically significant difference between the two groups or between Grades 10 and 11 in each group regarding the interaction time. In Grade 11, the interaction time was slightly less than in Grade 10. Interaction times in early–full and late–partial EMI schools were similar. Most of the interactions consisted of teacher talk which was at nearly 63% (Table 4.3), with a small percentage of student talk in both types of schools. Yet, there was more student talk in the early–full EMI schools and less teacher talk in the late–partial EMI schools.

Similarly, there was no statistical difference in the average length of teacher talk and student talk in the two types of schools. However, teacher talk contributed to the majority of the interactions in the two types of schools. The results then showed that although there were more teacher and student turns in the late–partial EMI schools, the average length of those turns was shorter compared with the early–full EMI schools. The student turns were also much shorter (3.16 s < 8.48 s). The maximum length of student turns in the late–partial EMI schools was four times shorter (302.6 s < 1,192.1 s). However, this observation did not hold for Grade 11 when compared with Grade 10. There were more teacher turns and fewer student turns in Grade 11 in both types of EMI schools. The maximum length of Grade 11 student turns was shorter in both school groups.

4.3.2   L1/L2 use

Analyses showed that even though all participating schools claimed to be EMI schools, there was a disparity in the actual amount of English spoken in the classrooms. With regard to the percentages of L2 and L1 spoken, there was a statistically significant difference between the early–full EMI schools and the late–partial EMI schools with a large effect size. A high percentage of L2 was observed in the early–full EMI schools. A more balanced percentage of L1 and L2 was observed in the late–partial EMI schools. However, there was no statistically significant difference between Grades 10 and 11 in each school group. There was more L1 use and less L2 use in both types of schools, though the manner of code-switching varied.

Teachers and students in the early–full EMI schools used a high percentage of L2 during the interactions. With more L2 use (87%), the percentages of L2 use suggested that early–full EMI schools were genuine EMI schools. There was more L1 use in Grade 11 than in Grade 10. The transcriptions indicate a few occasions when teachers and students used L1 to supplement their English explanations, typically for a quick translation of technical vocabulary or to provide examples of daily experiences related to science. In late–partial EMI schools, there was a similar ratio of L1 and L2 use between teachers and students. There was a slight increase in L1 use among Grade 11 students than those in Grade 10 (46% < 54%).

To examine code-switching, I measured two types of code-switching patterns: inter-sentential and intra-sentential. A statistical difference was found for the intra-sentential code-switching between early–full and late–partial EMI schools. The data showed that early–full EMI schools had more inter-sentential code-switching (between sentences) (61% > 49%), suggesting that teachers and students with more experience in EMI instruction produce more complete sentences in English. They used English as the matrix language, embedded with Cantonese translations. However, the late–partial EMI schools used more intra-sentential code-switching, suggesting that teachers and students tended to use Cantonese as the matrix language, embedded with English technical terms. Teachers and students also tended to interact socially about non-academic items in complete Cantonese discourse.

4.3.3 Pedagogical functions of the interactions

The results showed that the four most frequent pedagogical functions in the interactional data were similar in both types of EMI schools. The ranking was: teachers' informs, teachers' self-utterances, teachers' directs, and teachers' replies. The fifth most frequent pedagogical function was different in the two types of schools: 'check' (early–full EMI) and 'express opinion' (late–partial EMI). In the early–full EMI schools, there was more 'check' than 'self-utterance' in Grade 11, but this pattern was reversed in Grade 10. In the late–partial EMI schools, there was more 'self-utterance' than 'direct' in Grade 11, but the pattern was reversed yet again in Grade 10.

4.3.4 Teacher's feedback and wait time

There was no statistically significant difference between early–full and late–partial EMI schools in the percentages of teacher feedback. The only statistically significant difference was the number of affirmation-direct instruction feedback moves between Grades 10 and 11 in the late–partial EMI schools. Both types of schools had a similar distribution of teacher feedback. The frequency order was: 'affirmation-direct instruction' > 'extension by responsive questioning: focusing and zooming' > 'explicit correction-direct instruction' > 'constructive challenge'. From Grades 10 to 11, more 'affirmation-direct instruction' was observed in the early–full EMI schools but this pattern was reversed in the late–partial EMI schools.

Although there was no statistically significant difference between the two types of schools and across grades in their percentages of wait time, there was nonetheless a difference. In the late–partial EMI schools, the wait time was much longer than in the early–full EMI schools. From Grades 10 to 11, there was a shorter wait time in the early–full EMI schools, but a longer wait time in the late–partial EMI schools.

Interaction time and the distribution of teacher and student talk were similar in early–full and late–partial EMI schools. In short, 62–68% of the interactions were teacher talk, with very little student talk (5–8%), and the average length of teachers' turns was longer than students' turns. Teachers in both types of schools used a similar set of pedagogical functions during interaction (e.g., teachers' informs, teachers' self-utterances, teachers' directs, and teachers' replies) and offered similar types of feedback (e.g., 'affirmation-direct instruction' > 'extension by responsive questioning: focusing and zooming' > 'explicit correction-direct instruction' > 'constructive challenge').

However, the similarity in the interactions between early–full and late–partial EMI schools was limited to the macro level. When I examined the interactions at the micro level in terms of the average length of teacher and student talk, the ratio between L1 and L2 use, the patterns of code-switching (inter- and intra-sentential), and the types of teacher questions and wait time, several subtle differences across the two types of schools were found. In the early–full EMI schools, the average length of teacher talks and student talks was greater. Teachers and students used more L2 than L1 in their interactions, with the majority of interactions spoken in English. Teachers and students tended to code-switch between English and Cantonese at sentence level (inter-sentential). Teachers used a larger number of lower-order questions to recall students' previous learning of science concepts, and thus the wait time between teachers' initiations and students' responses was shorter.

On the other hand, in the late–partial EMI schools, teacher and student talk was shorter. For example, the maximum length of student talk was four times shorter than that in the early–full schools. Teachers used a greater number of higher-order questions to prompt students to use higher cognitive skills, and thus the wait time was much longer than in the early–full EMI schools. There was a tendency to use higher-order questions in Grade 11. Teachers and students used less L2 in interactions, with a similar ratio between L1 and L2 use in both teacher and student talk. Teachers and students tended to insert English technical terms into Cantonese interactions.

In terms of comparisons between Grades 10 and 11, both types of schools share similarities. For instance, similar patterns were observed in L1 and L2 use. More code-switching between sentences (inter-sentential) was observed in Grade 11 in both types of schools. However, there were several differences between the two types of schools. In the early–full EMI schools, the difference in the mean length of teacher turns was statistically significant; there was more teacher talk but less student talk in the early–full EMI schools. In the late–partial EMI schools, the average length of teacher and student turns was much less than in the early–full EMI schools, with the addition of more L1 use.

When I examined the pedagogical functions of the interactions, the function of interactive moves and the choices of teachers' questions were the two key indicators in marking the differences between early–full and late–partial EMI schools. The four most frequent pedagogical functions in the interaction sequences—inform, self-utterance, direct and reply—appear to reflect the nature of science teaching in Hong Kong. They suggest classrooms that are teacher-centred and authoritative, with very few opportunities for students to negotiate meaning with teachers at the lexical level. These subtle differences suggest that the interactions in early–full and late–partial EMI schools were different, which will be further explained in the discussion chapter.

4.4 Summary

The statistics detailed in this chapter indicate that early–full and late–partial EMI science classrooms are characterised by similar interaction patterns (percentages of interaction time, distribution of time between teacher and student talk, frequency of pedagogical functions) but the nature of the interactions is different (use of code-mixing, proportion of L1 and L2 use).

Key findings include the greater amount of L2 used in the early–full schools than in the late–partial schools, the greater use of inter-sentential code-mixing in the early–full schools, the greater use of intra-sentential code-mixing in the late–partial ones, and the greater use of higher-order questions in the late–partial schools than in the early–full schools. Particularly, in late–partial EMI schools, there were more student initiations and responses (though shorter), with more higher-order questions from the teachers in Cantonese but less direct feedback to students. Both teachers and students from both types of schools tended to use more L1.

In both types of schools, there was less interaction time in Grade 11 than in Grade 10. Additionally, there was more use of L1 in Grade 11 than in Grade 10. This is presumably because the Grade 11 syllabus is more cognitively challenging, and so code-switching is required more often. The maximum length of student turns in Grade 11 was also less than in Grade 10.

The following chapter will further examine the subtle differences in classroom interactions between the two types of EMI schools. Drawing on a case study of four EMI classroom transcripts, Chapter 5 will explore the nature of classroom interactions qualitatively and identify the features of language used in EMI science classrooms. The analysis will then illustrate how these interactions are similar and different in teaching science between early–full and late–partial EMI schools and the pedagogies implied by these interactive sequences.

Notes

1 Average lesson length is the sum of all the recording time of each lesson divided by the total number of lessons in each type of school, e.g., early–full EMI school = the sum of recording time of the 10 lessons in early–full EMI school divided by 10.
2 Average interaction time per lesson is the sum of teachers' and students' talk in all observed lessons divided by the number of lessons in each school type.
3 Percentage of interaction is the total interaction time divided by the total lesson time.
4 r (effect size) = z / square root of N, where N = total number of cases.
5 Teacher talk is the time taken up by teachers' utterances in an observed lesson.
6 Student talk is the time taken up by students' utterances in an observed lesson.
7 Percentage of interaction is total interaction time divided by total lesson time.
8 r (effect size) = z / square root of N, where N = total number of cases.

9 The percentage of the science teacher's feedback is calculated by dividing the number of each type of feedback for a teacher divided by the total of all different types of feedback by that teacher.

10 Wait time is defined as the pauses between a teacher finishing speaking and a student starting a turn, and the gap between a student finishing a turn and teacher starting one (Ingram & Elliott, 2014).

11 The percentage of L1 use is the number of turns in L1 divided by the total number of turns. The percentage of L2 use is the number of turns in L2 divided by the total number of turns. The aim of this counting was to see how frequently teachers and students use L1 or L2 in each turn. Given that the differences between English and Cantonese create challenges for calculating actual use by taking into account the length of those turns, or by the number of syllables/words, I only performed the calculation based on the number of turns. Some teachers and students code-switched at different levels, some with a full sentence, some with a noun phrase, or some with a single grammatical item such as a verb or adverb. This made it difficult to perform the calculation based on the length of turns or the number of words or syllables.

12 r (effect size) = z / square root of N, where N = total number of cases.

References

Chin, C. (2006). Classroom interactions in science: Teacher questioning and feedback to students' responses. *International Journal of Science Education, 28*(11), 1315–1346. http://doi.org/10.1080/09500690600621100

Cohen, J. (1988). *Statistical power analysis for the behavioural sciences*. New York: Lawrence Earlbaum Associates.

Lo, Y. Y., & Macaro, E. (2012). The medium of instruction and classroom interaction: Evidence from Hong Kong secondary schools. *International Journal of Bilingual Education and Bilingualism, 15*(1), 29–52. http://doi.org/10.1080/13670050.2011.588307

Tsui, A. B. M. (1985). Analysing input and interaction in second language classrooms. *RELC Journal, 16*(1), 8–32.

5 Qualitative analysis of classroom interactions in early–full and late–partial EMI science classrooms

This chapter presents a picture of science pedagogies and classroom interactions in four biology lessons. Two of the lessons are from early–full English medium instruction (EMI) schools and two are from late–partial EMI schools, with one Grade 10 lesson and one Grade 11 lesson from each type of school. The transcripts from these schools were chosen because the lessons covered similar science topics and therefore provided a homogeneous dataset for exploring the nature of classroom interactions. This chapter includes a descriptive analysis of science pedagogy using Mortimer and Scott's (2003) framework, which addresses the teaching purpose, communicative approaches, patterns of interaction, and forms of interaction in each of the four lessons. It also adopts the frameworks of Fang (2006) and Kong and Hoare (2012) for identifying language choices made in the classrooms. The chapter concludes with an analysis of interactive sequences in the data to demonstrate how an EMI science teacher interacted with the students to develop their thematic understanding of science in two of the biology lessons. In all, the analysis in this chapter may provide insights into the nature of interaction in EMI science classrooms in Hong Kong secondary schools.

5.1 An integrated framework for analysing classroom interaction qualitatively

As explained in Chapter 3, the integrated framework I am drawing on in this analysis highlights the three dimensions relevant to classroom interactions: science pedagogy, the language of science, and the understanding of science concepts. Figure 5.1 illustrates this framework.

In the case of science pedagogy, the purpose of the analysis is to understand how teachers and students achieve the purpose of learning science in classroom interactions. Mortimer and Scott (2003) provide an analytical framework that categorises classroom interactions in relation to science pedagogy: teaching purpose, teaching content, communicative approach (non-interactive versus interactive; authoritative versus dialogic), teachers' interventions, the pattern of interactions, and the form of interactions.

DOI: 10.4324/9781003001454-6

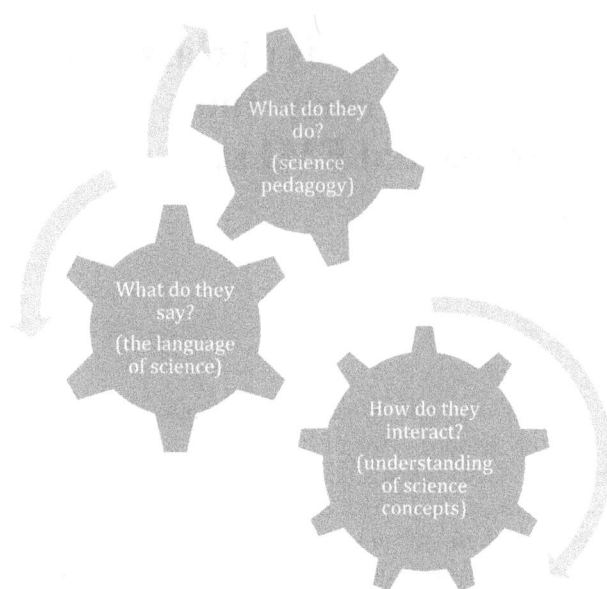

Figure 5.1 Conceptual framework of the three elements to analyse classroom interactions.

To describe the language of science, Fang's (2006) and Kong and Hoare's (2012) frameworks provide a method for analysing the language features the teachers and students use in EMI science classrooms. Kong and Hoare examined teachers' talk and classroom interactions in EMI content–subject classrooms. They conducted a detailed discourse analysis of classroom data (teacher and student talk transcripts, teacher interviews, direct observation, lesson plans, and teaching materials) from an EMI secondary science classroom in Hong Kong and provided a framework for analysis accordingly. Like Kong and Hoare's study, this study explores how teachers develop students' academic language skills and analyses the specific linguistic features, such as the use of nominalisations (e.g., nominal phrases), that students use when constructing academic knowledge in the EMI classroom. Drawing from Fang's study (2006), I provide a comprehensive review of the specific linguistic features that high school science students encounter when learning science through English.

To capture how students develop an understanding of science concepts through interaction, I based my analysis on Lemke's (1990) concept of semantic relations. Since the primary goal of the content classroom is to develop students' content knowledge, it is especially important to understand how science learning occurs through interaction. Lemke's (1990) concept of semantic relations provides a framework for describing how science teachers modify their talk to provide comprehensible input that develops students' understanding of science through interaction.

Table 5.1 summarises the three analytical aspects of the integrated framework. By using this three-layered integrative framework, I was able to understand how

Table 5.1 The three-layer integrated framework used in this study

Aspects of classroom interaction examined	Theoretical frameworks	Aspects of analysis for classroom interaction
Science pedagogy	Mortimer and Scott's framework (2003)	Analysis of the science pedagogy (teaching topic, communicative approach, patterns of interaction (IRF), teachers' interventions)
The language of science	Kong and Hoare (2012)	The use of academic language through teacher and student talk in science classrooms (lexical density, subject-specific vocabulary, grammatical metaphors, complex noun phrases, language of knowledge relationship)
	Fang (2006)	The language of science in teachers' talk and interaction with students (vocabulary, grammar, forms)
Thematic development in science	Lemke (1990)	How teachers develop students' understanding of technicality and logical reasoning in science through interactions. (Unpacking nominalisation, making connections between different taxonomic knowledge)

science teachers interact differently with their students in EMI classrooms and how teachers modify their talk to provide comprehensible input to help students develop their understanding of science concepts through interaction.

The following discussion presents the qualitative analysis of classroom interactions for each of the three different dimensions in turn.

5.2   Analysis of classroom interactions in four biology lessons

5.2.1   *Early–full EMI Biology Grade 10*

This lesson extract is from a Grade 10 early–full EMI biology classroom. The students have been using EMI for three years and have just started to study the senior science curriculum.

In this lesson, the teacher introduced students to the concept of photosynthesis. Specifically, he explored with students the possible factors that affect the rate of photosynthesis through a lively but systematically structured interaction. The interaction is described in the form of IRF sequences. As Excerpt 1A illustrates, there are five IRF sequences in the 20 turns, two of which are the extended IRF sequences. The teacher initiated the interaction by asking students to suggest possible

Excerpt 1A The lesson transcript of early–full EMI Grade 10 EMI Biology

Topic: Photosynthesis

Sub-topic: Factors that affect the rate of photosynthesis (light intensity, concentration of CO_2, oxygen, amount of chlorophyll, source of water)

Medium of instruction (MOI): English only

Turn	Speaker	Text	IRF sequence	Science topic
1	T	Can you tell me what sort of factors may also affect photosynthesis? What sort of factors may also affect photosynthesis?	I (Open Q)	Photosynthesis
2	S1	Amount of chlorophyll.	R	Amount of chlorophyll
3	T	Amount of chlorophyll, yes. Okay?	F (acknowledge)	
4	T	Amount of chlorophyll will have an effect. And besides that? Remember the equation of photosynthesis. You have CO_2, and you have what? CO_2 plus what?	I (open Q)	Equation of photosynthesis; carbon dioxide (CO_2)
5	S1	Water.	R	Water
6	T	Water, okay. Addition of water.	F (acknowledge)	
7	T	Together with the chlorophyll and also what sort of things?	I (open Q)	
8	S1	Light energy.	R	Light energy
9	T	Light energy. What is the product of the photosynthesis? Oxygen together with what?	F (ignore)	
10	S1	Starch or food.	R'	Starch
11	T	Okay, so light we've mentioned here already.	F' (ignore)	
12	T	And your classmate talked about whether it is due to the limit of chlorophyll. Any more things that limit photosynthesis?	I (open Q)	
13	S1	Water.	R	
14	T	Water, yes. Can you tell me the source of water of the plants? Where does it come from? Where does it come from?	F (acknowledge +expand)	Source of water

(Continued)

Excerpt 1A (Continued)

Turn	Speaker	Text	IRF sequence	Science topic
15	S1	Soil.	R'	Soil
16	T	Soil, yes. So, if inside the soil, the amount of water in the soil is very low, then it will be a very great limiting factor.	F' (acknowledge + expand)	
17	T	Anymore? Very straightforward, we have mentioned chlorophyll, we have mentioned water. And the last one should be...?	I (open Q)	
18	S1	CO_2.	R	
19	T	Why is CO_2 a limiting factor?	F (expand)	

factors that affect the rate of photosynthesis. Students gave their answers and the teacher evaluated their responses. If the answer was correct, the teacher affirmed the student's answer and moved on to the next factor. If the answer was inconsistent with the teacher's view, the teacher provided extensive feedback to the student. The IRF sequence was then repeated as the teacher continued to explore with the students how these factors (light intensity, concentration of CO_2, oxygen, amount of chlorophyll, source of water) affect the rate of photosynthesis.

Excerpt 1A shows that the teacher also used several teacher interventions (Mortimer & Scott, 2003) in the form of extended IRF sequences. These interventions included (1) focusing on a particular student's idea by zooming into a particular student's response, (2) expanding on what a student has said, (3) further exploring the ideas, (4) asking for clarification of student ideas, and (5) reviewing student responses with the whole class.

For example, in turn 19 (see Excerpt 1B), the teacher asked the students to elaborate on why carbon dioxide is a limiting factor for photosynthesis. Then, from turn 20 to turn 25, the teacher built on the students' answers to discuss the role of stomata in the process of gas exchange. In this period of interaction, the teacher intervened by developing the student's understanding of the scientific story. The teacher paid attention to a particular student's response (*stomata* in turn 20), and then marked the key idea by enacting a confirmatory exchange with the student (*because it comes from the stomata* in turn 23). Afterwards, the teacher checked the students' understanding by asking for clarification about the role of the stomata in plants (in turn 23). Finally, the teacher returned to and revised the idea again (in turn 25) by summarising the actual function of stomata in gas exchange.

In Excerpt 1B, it is evident that the teacher made several interventions from turns 20 to 25, including sharpening ideas, selecting ideas, marking key ideas, sharing ideas, checking student understanding, and reviewing. These efforts appear designed to help the students develop scientific reasoning by making it explicit that stomata are a controlling factor in the amount of carbon dioxide intake during gas exchange. This may have helped the students to understand the reason why carbon dioxide affects the rate of photosynthesis. Figure 5.2 summarises the number of turns and turn types in the complete lesson.

We can see that in this example of an early–full EMI Grade 10 biology lesson, there were 268 turns by teachers (173 turns) and students (95 turns) and 48 IRF sequences: 26 were extended IRF, and 22 were simple IRF. There were more open questions (41) than closed questions (7), and students provided 95 responses while the teacher made 94 feedback moves. A relatively high ratio of IRF[1] (18%) in the interaction, with more extended IRF sequences, suggests that there were opportunities for the students to interact and respond, and for the teacher to give feedback to the students.

In both Excerpts 1A and 1B, the teacher used a range of teacher interventions to develop students' understanding of the factors that affect photosynthesis. In

Excerpt 1B Lesson transcript of early–full EMI Grade 10 EMI biology

Turn	Speaker	Text	IRF sequence	Science topic
17	T	Anymore? Very straightforward, we have mentioned chlorophyll, we have mentioned water. And the last one should be…?	I (open Q)	Chlorophyll
18	S1	CO_2	R	Carbon dioxide (CO_2)
19	T	Why is CO_2 a limiting factor?	F (expand)	
20	S1	Stomata.	R'	
21	T	Limited by stomata. Why will it be limited by stomata? Can you tell me?	F' (acknowledge + expand)	Stomata
22	S1	Because it comes from the stomata.	R''	
23	T	Because it comes from the stomata. What is the role of the stomata in plants?	F'' (acknowledge + expand)	
24	S1	Gas exchange.	R'''	Gas exchange
25	T	Yes, it is the site of gas exchange. They allow the CO_2 from outside to go into the leaves. So, the number of stomata will have an effect. It will be one of the possible reasons.	F''' (acknowledge + expand)	

early–full G10 biology

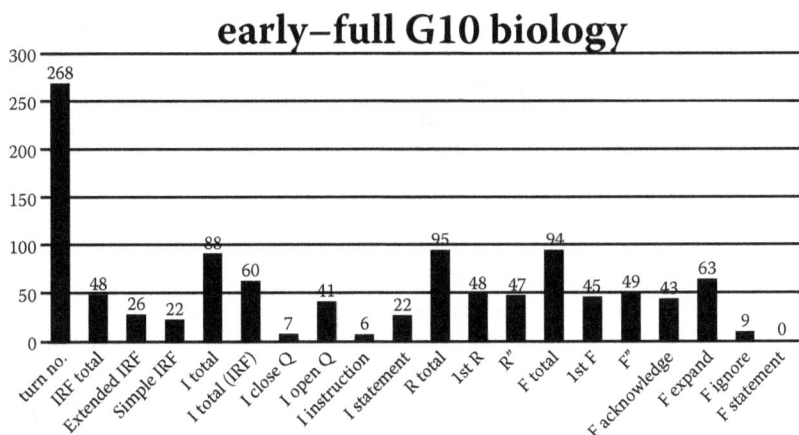

Figure 5.2 The number of turns, IRF in the whole lesson.

Excerpt 1C Science topics in excerpt 1a and 1b

Turn	Speaker	Text	IRF sequence
1	T	*... **factors** may also affect photosynthesis?*	I
2	S	**Amount of chlorophyll.**	R
3	T	*Amount of chlorophyll will have an effect, and besides that?*	F (acknowledge)
5	S	**Water.**	R
6	T	**Addition of water.**	F (acknowledge)
7	T	*Together with the chlorophyll and what sort of things?*	I (open Q)
8	S	**Light energy.**	R
9	T	*... **product of photosynthesis?***	F (ignore)
10	S	**Starch or food.**	R'
12	T	*Any more things that limit it?*	I (open Q)
13	S	**Water.**	R
14	T	Where does it come from?	F (expand)
15	S	**Soil.**	R'
16	T	Anymore? We have mentioned chlorophyll, we have mentioned water. *And the last one should be...?*	F' (acknowledge)
18	S	CO_2	R
19	T	Why is CO_2 a limiting factor?	F (expand)

Excerpt 1A, the teacher took account of his students' points of view on the factors. In Excerpt 1B, the teacher then explored those different views with his students to co-construct an understanding of how carbon dioxide is a limiting factor for photosynthesis.

Excerpt 1C below is a simplified version of Excerpt 1 that highlights the science topics discussed in the interaction in bold. The excerpt shows that the teacher made attempts to elicit students' further responses on the possible factors of photosynthesis by asking questions in his feedback moves. After the students answered topics A, B, and C, the teacher worked with the students to explore different views, typically by responding and providing feedback to each student's response, to generate new meanings and co-structure understandings of different views about the scientific phenomenon.

The following excerpt summarises the kind of language students used in classroom interactions. Students predominately produced incomplete sentences consisting of nouns or noun phrases referring to scientific items when responding to their teacher. In these exchanges, the teacher did not encourage students to produce complete sentences or make any attempts to correct any students' language mistakes. There was also no feedback provided to students on how to acquire the language of science in the L2 context (Figure 5.3; Excerpt 1D).

This example of an early–full EMI Grade 10 EMI biology lesson can be linked to Mortimer and Scott's (2003) framework as an example of an interactive and dialogic science lesson, as it contains several long sequences consisting of rapid turn-taking and plenty of questions and feedback from the teacher. Excerpts 1A and 1B are highly interactive, with both IRF and extended IRF sequences. The students' answers were long and their ideas were acknowledged by the teacher, which were then followed up by the teacher's feedback. Realised by open questions, the

Excerpt 1D Students' language output in excerpt 1

Turn	Speaker	Text	IRF sequence	Turn
2	S	Amount of chlorophyll	Noun phrase	R
5	S	Water	Noun	R
8	S	Light energy	Noun	R
10	S	Starch or food	Noun	R'
13	S	Water	Noun	R
15	S	Soil	Noun	R'
18	S	CO_2	Noun (technical vocabulary)	R

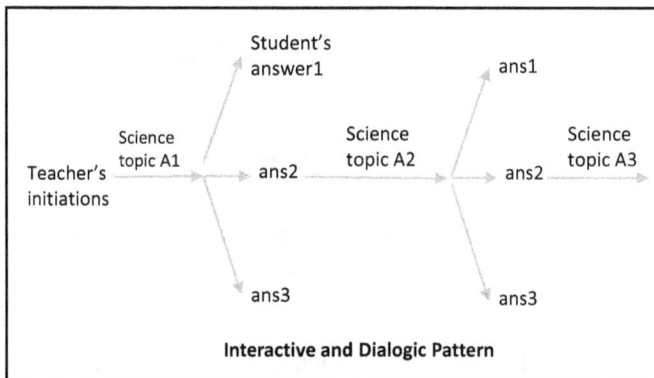

Figure 5.3 Conceptual diagram of the interactive and dialogic pattern.

teacher-initiated discussions frequently to facilitate extended student answers, which in turn enabled the teacher to explore the meaning and encourage students to practise their English. In this process, the teacher extended scientific ideas to students to help them understand abstract scientific knowledge. The teacher discussed with students what had been explored in the lesson, and this interaction provided opportunities for students to further study the content and practise or learn the language.

5.2.2   Early–full EMI Grade 11 biology

I now turn to the classroom interactions of 11th graders at the same school. These students have been in EMI for four years and have one more year to study senior-level science. The extract from this lesson comes from a biology classroom in an early–full EMI class. In this lesson, the teacher introduced the concept of cell division and checked the students' understanding of bacterial cell division at a microbiological level.

Excerpt 2A shows a considerable amount of interaction between the teacher and his students as the teacher checked the students' understanding of key concepts related to cell division (e.g., bacteria, fungi, yeast, genetic variation). There are two

Excerpt 2A Lesson transcript of early–full EMI Grade 11 EMI biology
Topic: Cell division; Sub-topic: Mitotic cell division; MOI: English only

Turn	Speaker	Text	IRF sequence	Science topic
1	T	What do you think? What microorganisms can also grow here? Apart from bacteria. You have learnt.	I (open Q)	Microorganisms; bacteria
2	S1	Fungi.	R	Fungi
3	T	Fungi? Yes, can you give me an example of fungi? You have learnt a very small, okay, unicellular fungi. What's that?	F (acknowledge + expand)	Unicellular fungi
4	S1	Yeast.	R"	Yeast
5	T	Yeast, yes. We can grow yeast here okay? This microorganism, this colony, is derived from the division, the repeated cell division of one cell okay, into many cells by mitotic cell division. It is the asexual reproduction of the bacteria. And the bacterial colony almost contains this number of bacteria, this amount that means about millions of that bacteria, okay, in this colony. So that you can know directly, use your naked eye, to observe this, okay?	F" (statements)	Colony; asexual reproduction; mitotic cell division

(Continued)

Excerpt 2A (Continued)

Turn	Speaker	Text	IRF sequence	Science topic
6	T	And the genetic materials of the bacteria in this colony should be the same. Why?	I (open Q)	Genetic materials of the bacteria in agar plate
7	S1	Because of mitotic cell division.	R	
8	T	Why does mitotic cell division produce the same genetic material?	F (acknowledge + expand)	
9	S1	No genetic variation.	R"	Genetic variation
10	T	No genetic variation. Okay? Genetic variation is very important this term. No genetic variation in mitotic cell division, okay?	F" (acknowledge)	
11	T	Any questions about this petri dish? Any questions?	I (close Q)	Petri dish
12	S1	No.	R	

extended IRF sequences in the 12 turns. In these IRF sequences, the teacher first asked students what microorganisms are present, prompting them to answer (e.g., fungi). The teacher then asked students to give examples (e.g., yeast). Instead of probing students' answers, the teacher explained how yeast is related to cell division (in turn 5). The teacher then jumped to another topic about genetic material and asked students to give a reason why the genetic material is the same as in a population of cells (in turn 6). Students answered quickly, saying that it was because of mitosis in the cells. The teacher then pressed for a reason and the students responded that there was no genetic variation. In these exchanges, we did not see the teacher expand on the students' answers or provide any feedback to the students.

As shown in Figure 5.4, in this example of an early–fully EMI Grade 11 biology lesson, the teacher and students took 176 turns with a total of 45 IRF sequences: 16 were extended IRFs and 29 were simple IRFs. The higher proportion of IRFs in the interactions (26%) and the simple IRF sequences indicate fewer opportunities for students to answer and for the teacher to provide feedback to the students. There were more open-ended questions (28) than closed-ended questions (10). There were 47 student responses and 43 instances of teacher feedback moves.

Excerpt 2B is a simplified version of Excerpt 2A, highlighting the science topics discussed during the interaction. In turn 1, the teacher began by asking the students a question to remind them what they learnt in the previous lesson. From turns 2 to 5, the students said 'fungi' which was not the answer the teacher expected. Instead of addressing the students' answers, the teacher continued to provide more prompts to help the students come up with the answer 'yeast'. Once the students said 'yeast', the teacher started to elaborate in a sustained turn (5). Similarly, in turn 6, the teacher asked the students to explain why the genetic materials are the

early–full G11

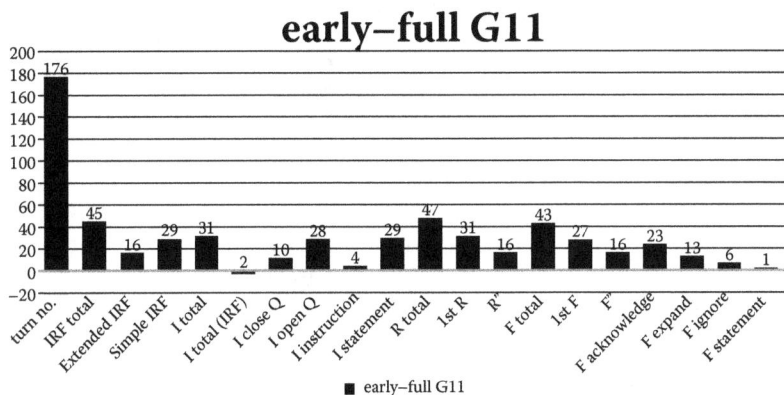

Figure 5.4 The number of turns and IRF sequences in the lesson.

Excerpt 2B Science topics in excerpt 2a

Turn	Speaker	Text	IRF sequence
1	T	What do you think? What microorganisms can also grow here? Apart from bacteria. You have learnt.	I (open Q)
2	S1	**Fungi**	R
3	T	**Fungi?** You have learnt a very small, okay, unicellular fungi. What's that?	F (expand)
4	S1	**Yeast.**	R"
5	T	**Yeast,** yes. We can grow yeast here okay? This microorganism...	F" (statements)
6	T	And the **genetic materials** of the bacteria in this colony should be the same. Why?	I (open Q)
7	S1	Because of **mitotic cell division.**	R
8	T	Why does mitotic cell division produce the same genetic material?	F (expand)
9	S1	No genetic variation.	R"
10	T	No genetic variation. Okay? **Genetic variation is very important** this term. No genetic variation in mitotic cell division, okay?	F" (acknowledge)

same as the colony of cells. The teacher did not address the students' answer in turn 7 until the students said 'no genetic variation' in turn 9. The teacher responded by claiming that there is no genetic variation in cell division.

Excerpt 2C summarises the types of language used by students in classroom interactions. It shows that students in Grade 11 were similar to the Grade 10 students in producing a number of incomplete sentences consisting of nouns or noun phrases referring to scientific items when responding to their teacher. In these exchanges, the teacher did not encourage the students to produce a complete sentence, attempt

Excerpt 2C Students' language output in excerpt 2a

Turn	Speaker	Text	Grammatical categories	IRF sequence
2	S1	Fungi	Noun (technical)	R
4	S1	Yeast	Noun (technical)	R"
7	S1	Because of mitotic cell division	Noun phrase (technical)	R
9	S1	No genetic variation	Noun phrase (technical)	R"
12	S1	No	Discourse marker	R

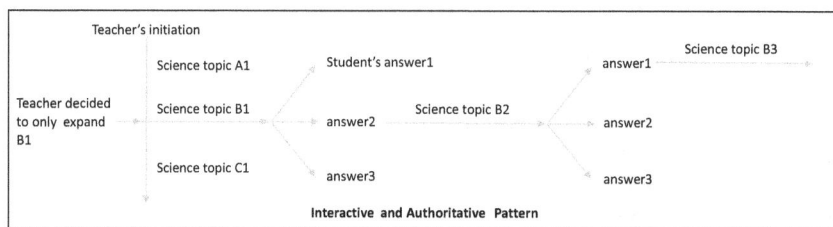

Figure 5.5 Conceptual diagram of the interactive and authoritative pattern.

to correct the students' linguistic errors or provide any feedback on how to master the language of science in an L2 context.

In these examples, after the student gave different answers for science topics (fungi, yeast, mitotic cell division and genetic variation), the teacher decided which of the answers was consistent with his planned science explanation and focused on it by asking more questions.

Concerning Mortimer and Scott's (2003) framework, the transcript of early–full EMI Grade 10 EMI biology is an example of an interactive and authoritative science lesson (Figure 5.5).

This lesson was characterised by longer teacher-initiated periods and more statements, demonstrating the power of the teacher to control the classroom. However, there was still a considerable amount of teacher–student interaction in the form of IRF sequences. There were many instances of teacher initiation through closed questions or statements that seemed to be aimed at simply stating the teacher's view of the planned science topic or checking for student understanding, with brief student responses. There were some forms of student responses followed by feedback from the teacher affirming that the students' responses were consistent with the teacher's planned explanations. Although teachers sometimes considered students' answers, none of these students' ideas were further explored or followed up. During the interactions, the teacher merely articulated abstract scientific knowledge. The teacher asked students for their ideas without engaging much with the students' explanations of the topic. Students had the opportunity to learn the content but had limited opportunity to practice the language.

5.2.3 Late–partial Grade 10 biology

In this section, I present data from students with little or no experience with EMI at the same secondary level. The following Excerpt 3A is taken from a mid- to late Grade 10 biology classroom. In this class, the teacher introduces the concept of immunity and helps students distinguish between specific and general immune responses. Specifically, the teacher asked students to explain the specific immune response.

Extract 3A shows a significant amount of interaction between the teacher and students in the form of IRF sequences exploring their understanding of specific immune responses. This was achieved through the extensive use of Cantonese throughout the interaction. Teachers were able to explain difficult scientific concepts, and ask higher-order questions (e.g., in the second round, 'Why do we call these specific?'). Students were able to understand the teacher's higher-order

Excerpt 3A The lesson transcript of late–partial Grade 10 EMI Biology

Topic: Immunity; Sub-topic: Specific immune response
MOI: Cantonese as the matrix language, embedded with English vocabulary
(*Italics: Cantonese;* non-italics: English)

Turn	Speaker	Text	IRF sequence	Science topic
1	T	*First, we are talking about the* 首先，我地會講概係有關 specific immune response; *this is not* 這個唔係 a general immune response. *It is* a 這個係 specific immune response. *When we talk about* 當我地講 skin, saliva, tears, gastric juice, *we call these* 我地講這些叫 general.	I (statement)	Specific immune response
2	T	*Why do we call these* 點解 我地叫 specific? *Anyone remember* 有無人記得? *Why would these be regarded as* 點解這些叫 specific?	I (open Q)	
3	T	*Previously we talked about* 之前我地講 inflammation *and* 同 blood clotting.	I (statement)	Inflammation, blood clotting
4	T	*Why do we call these* 點解我地 叫 non-specific? *Why would we call them like this* 點解我 地咁樣叫? *This would be* 這 樣叫 specific? 什麼叫 *What is* specific?	I (open Q)	Non-specific
6	T	Target. *That's right* 對呀. Targeting the **pathogens.**	F (expand)	**Pathogens**

(*Continued*)

Excerpt 3A (Continued)

Turn	Speaker	Text	IRF sequence	Science topic
7	T	*Why would we say that it has the ability of* 點解我地講佢有能力去 targeting? *You can find a very important word here* 你可以在這找到重要的字。	I (open Q)	
8	S2	**Antigen**	R	
9	T	**Antigen**. *What does it mean?* 點解? *You need to know what it means rather than copy it* 你需要知道點解重要過抄寫.	F (expand)	**Antigen**
10	T	Whose antigen belongs to? *Does it belong to our body or* 係唔係屬於身體定 pathogen?	I (open Q)	
11	S3	**Pathogen.**	R	Pathogen
12	T	An antigen is a type of protein.	F (ignore)	
13	T	*If the* 如果 pathogen *has this type of* 係有這種 antigen, 這種 *this type of* protein, **leukocytes** 係我地身體入面會認得這些野唔屬於身體 *inside our body would recognise that these are not something belonging to the body.* 當佢認到這個唔係 *When it can identify that this is not some* substance 屬於身體 *belonging to the body, this* 佢 antigen 就會當作認到 *would be regarded as recognised. When I say* 當我講 non-specific, 當 *when the* skin 要拒絕唔俾 *needs to resist, not allowing the* bacteria 或者外來物質進入身體 *or foreign substances to come inside the body,*	I (statement)	**Leukocytes**; non-specific
14	T	*will the* 咁 skin 會認得邊一種物質同唔俾進入 *recognise which substance and not allow its entry*?	I (close Q)	
15	S1	No.	R	
16	T	*So it is* 所以就叫 non-specific. *There is no* 佢無 recognition, 所有野入唔到 *everything cannot go in. No* 唔需要 recognition *is needed.*	F (acknowledge)	Recognition

questions and answered the questions in Cantonese. However, students' responses were limited and usually consisted of a single word rather than a complete sentence.

From turns 4 to 16, there were three simple IRF sequences in which the teacher asked students to describe the characteristics of a particular immune response. For example, various students stated the target (turn 5), antigen (turn 8), and pathogen (turn 11). Teachers provided feedback on students' responses in the form of open-ended questions. In question 13, the teacher summarised what he had discussed with students about 'pathogens and antigens' and established a cause-and-effect relationship between these two terms, i.e., 'a pathogen has this type of antigen'. Finally, he introduces a new term, 'leukocytes', the type of cells that recognise pathogens to initiate a specific immune response. The teacher did not make this connection explicit to the students.

As Figure 5.6 shows, in this example of late–partial Grade 10 EMI biology, the teacher and students took 232 turns, in 32 IRF sequences: 13 were extended IRF and 19 were simple IRF. There were more open questions (49) than closed questions (7), and more teacher feedback moves (95) than student responses (67). There is a moderate ratio of IRF (14%) in the interaction with more extended IRF sequences, suggesting that there were opportunities for the students to respond and for the teacher to give feedback to students.

Excerpt 3B below is a simplified version of Excerpt 3A, highlighting the science topics discussed in the interaction. We can see that the teacher made attempts by asking questions in the feedback move to elicit students' further responses on the characteristics of specific immune responses from turns 4 to 11. Students provided answers (e.g., target, antigen, pathogen). In turn 13, the teacher summarised what he discussed with the students.

Meanwhile, Excerpt 3C summarises the kind of language students produced in classroom interaction. The data shows that students from late–partial schools are very similar to the students from early–full EMI schools, as seen in previous

late–partial G10 biology lesson

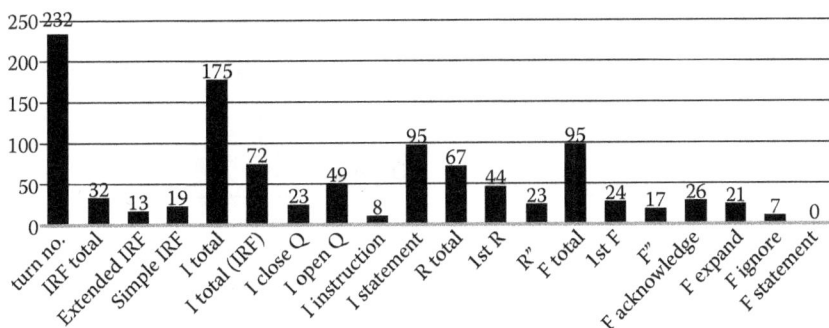

Figure 5.6 The number of turns and IRF sequences in the whole lesson.

Excerpt 3B Science topics in excerpt 3A

Turn	Speaker	Science topics	IRF sequence
4	T	*Why do we call these* non-specific? *Why do we call them like this? This would be specific? What is* specific?	I (open Q)
5	S1	**Target.**	R
6	T	Target. *That's right.* Targeting the **pathogens.**	F (expand)
7	T	*Why do we say that it has the ability of targeting?*	I (open Q)
8	S2	**Antigen.**	R
9	T	**Antigen.** *What does it mean?*	F (expand)
10	T	Whose antigen belongs to? *Does it belong to our body or* pathogen?	I (open Q)
11	S3	**Pathogen.**	R
12	T	An antigen is a type of protein.	F (ignore)
13	T	*If the* pathogen *has this type of* antigen, *this type of* protein, *the* **leukocytes** *inside our body would recognise that these are not something belonging to the body. When it can identify that this is not some* substance *belonging to the body, this* antigen *would be regarded as recognised. When I say* non-specific, *when the* skin *needs to resist, it does not allow the* bacteria *or foreign substances to come inside the body,*	I (statement)
14	T	*Will the* skin *recognise which substance it is and not allow its entry?*	I (close Q)
15	S1	No.	R
16	T	*So it is* non-specific. *There is no* recognition, *everything cannot go in. No* recognition *is needed.*	F (acknowledge)

Excerpt 3C Students' language output in excerpt 3A

Turn	Speaker	Text	Grammatical categories	IRF sequence
5	S1	Target	Noun (technical)	R
8	S2	Antigen	Noun (technical)	R
11	S3	Pathogen	Noun (technical)	R
15	S1	No	Discourse marker	R

examples, in that they predominately produced incomplete sentences consisting of nouns or noun phrases referring to scientific items when responding to their teachers. However, students in this lesson, as seen in Excerpt 3A, could only respond to their teachers using the technical items mentioned in previous interactions. In Excerpt 3A, the teacher again did not encourage students to produce a

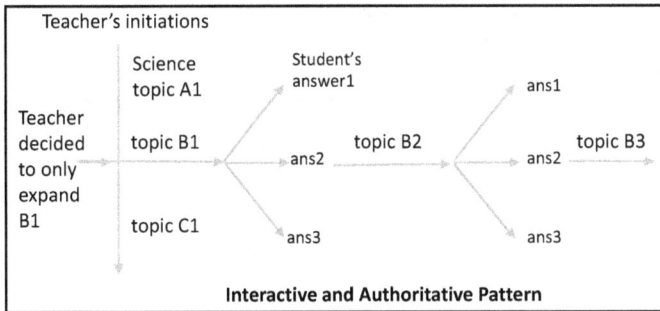

Figure 5.7 Conceptual diagram of the interactive and authoritative pattern no. 2.

complete sentence, make any attempt to correct students' language mistakes or provide any feedback to students about how to acquire the language of science in the L2 context.

Similar to the findings from Section 5.3.2, the teacher decided which of the students' answers was consistent with his planned science explanation, which he then focused on by asking more questions. Based on Mortimer and Scott's (2003) framework, the example in this section also falls in line with an interactive and authoritative science classroom (Figure 5.7).

5.2.4 *Late–partial Grade 11 biology*

Excerpt 4A is from a Grade 11 late–partial biology classroom, where the students have little EMI experience and have been studying the senior science curriculum for one year. In this lesson, the teacher introduced the concept of chromosomes. The teacher asked the students to guess the number of pairs of chromosomes in humans and the number of chromosomes in each pair from the father and mother. She used the analogy of counting postage stamps to help students understand the total sets and the number of chromosomes in a set. She then helped students to distinguish between diploid (two sets of chromosomes, 2n) and haploid cells (one set of chromosomes, 1n) in sexual reproduction.

In Excerpt 4A, there were three extended IRF sequences in 35 turns. One of the extended IRF sequences had more than three students' responses and teacher's feedback (from turn 191). There was a considerable amount of interaction between the teacher and students in the form of IRF sequences exploring students' under-standing of how to count the number of chromosomes in human cells. Across the interaction, the teacher used a number of technical terms (e.g., homologous chro-mosomes in turn 171; diploid cells and haploid cells in turn 183; paternal chromo-some and maternal chromosome in turn 191 and 193). On numerous occasions, the teacher used Cantonese (highlighted in italics in the transcript) to explain these ab-stract technical terms and to provide hints to help students understand these terms (e.g., in turn 195).

Excerpt 4A Lesson transcript of late–partial Grade 11 EMI biology

Topic: Immunity; Sub-topic: Specific immune response; MOI: Cantonese as the matrix language, embedded with English vocabulary (*Italics: Cantonese*; non-italics: English)

Turn	Speaker	Text	IRF sequence	Science topics
164	T	*How many* chromosomes *come from your* father's sperm?	I (open Q)	Father's sperm
165	S1	23.	R	
166	T	*How many come from your* mother?	I (open Q)	
167	S1	23.	R	
168	T	23. *So for* each pair, *why does one come from the* father *and one come from the* mother?	F (expand)	
169	T	*Because you are formed from* fertilisation, *i.e. the* fusion of the sperm and egg nuclei.	I (instatement)	Fusion of the sperm and egg nuclei.
170	T	*Okay, one comes from the father, the* paternal chromosome, *the blue one, for example here, and one comes from your mother,* the maternal chromosome.	I (instatement)	
171	T	*So in each pair, there are blue and red long* chromosomes, homologous chromosomes, homologous chromosomes. *That means they have the same size and same shape. One from the* father, *one from the* mother. Homo *means the same.*	I (instatement)	*Blue and red long chromosomes;* homologous chromosomes; 'Homo' means the same.
172	T	*Here, how many pairs are there? How many pairs of* homologous chromosomes *are there?*	I (open Q)	
173	S1	Two.	R	
174	T	Two pairs.	F (acknowledge)	
175	T	*You can see that each contains a special set. For example, homologous chromosome, we have stamps of 10 cents, 20 cents, $1.7, $2.4, this is one set. If we have two sets, we double it.*	I (statement)	
176	T	*How many sets are there?*	I (open Q)	
177	S1	*Two.*	R	

(*Continued*)

Excerpt 4A (Continued)

Turn	Speaker	Text	IRF sequence	Science topics
178	T	*Two. Our cells have two sets? How many sets do sperms and eggs have?*	F (expand)	
179	S1	*One set.*	R'	
180	T	*One for each type.*	F (acknowledge)	
181	T	*If there are 5 types of stamps, and I have one for each type. If I have two, I have two sets. Two sets, we call it* diploid cells. *If there is only one,* haploid.	I (statement)	Diploid cells; haploid
182	T	Diploid two, haploid one. *Okay?*	I (close Q)	
183	T	Diploid cells, haploid cells. *How many* haploid cells *are there in humans?*	I (open Q)	
184	S1	*Only two kinds.*	R	
185	T	*Which kind of* haploid cells?	F (expand)	
186	S1	Sperm and Egg.	R'	
187	T	*Other cells, skin, bones, hair, are all* diploid cells.	F' (ignore)	
188	T	Understand what I mean?	I (close Q)	
189	T	*Why is it called* diploid cells?	I (open Q)	
190	T	*Because when you find a* chromosome, *it always has a partner. Always in a pair. They are called* diploid cells.	I (statement)	*Partner*
191	T	*If it comes from the father, what is the name of the* chromosome?	I (open Q)	
192	S1	Paternal chromosome.	R	Paternal chromosome.
193	T	*Right, For mother?*	F (expand)	
194	S1	Maternal chromosome.	R'	
195	T	*Okay, For the same size* chromosomes *paired together, what are they called?*	F'(expand)	
196	S1	Homologous chromosomes	R"	
197	T	Homologous chromosomes. `	F"(acknowledge)	Homologous chromosomes
198	S1	Homologous chromosomes. `	R'"	
199	T	*That means the* chromosomes have the same shape and size, *or sometimes you say the same structure and size.*	F'"(ignore)	

As Figure 5.8 shows, in this late–partial Grade 11 EMI biology classroom, there were 185 turns taken between teacher and students and 19 IRF sequences: three were extended IRF and six were simple IRF. Similar to previous findings, there were more open questions (49) than closed questions (9). There were 28 student responses and 23 teacher feedback moves. There is a relatively low ratio of IRF (10%) in the interaction but with more extended IRF sequences, suggesting that there were fewer opportunities for the students to respond and the teacher to give feedback to the students.

Excerpt 4B is a simplified version of Excerpt 4A, highlighting (in bold) the science topics discussed in the interaction. We see that from turns 164 to 188, the teacher asked several questions about the number of chromosomes and the number of haploid cells and diploid cells. The teacher only responded to students' answers if they were consistent with the teacher's view. She only acknowledged what students said if their answer was similar to her views. There was no effort made to elicit students' further responses or to explore students' views. In turn 188, the teacher tried to transfer her understanding to her students by saying 'understand what I mean'. Then in turns 190 and 199, the teacher provided her own definition of diploid cells and homologous chromosomes.

Excerpt 4C, a summary of the students' language, shows that students predominately produced incomplete sentences consisting of nouns or noun phrases referring to scientific items. One noticeable difference in this example compared with the previous ones is that students produced technical English terms when responding to their teacher in Cantonese (L1). In the earlier examples presented above, teachers predominately used English for instruction. In these exchanges, the teacher again did not encourage students to produce a complete sentence make any attempts to correct any students' language mistakes or provide any feedback to students about how to acquire the language of science in the L2 context.

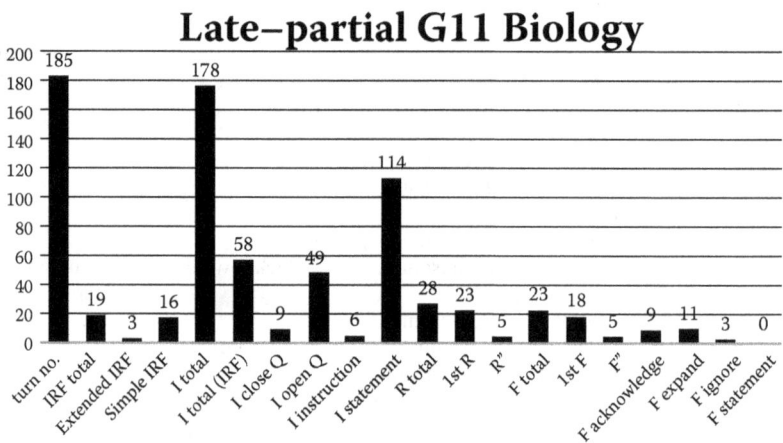

Figure 5.8 The number of turns and IRF sequences in the whole lesson.

Excerpt 4B The science topics in excerpt 4A

Turn	Speaker	Text	IRF sequence
164	T	*How many* **chromosomes** (father's sperm)?	I (open Q)
165	S1	23.	R
166	T	*How many* (mother's)?	I (open Q)
167	S1	23.	R
168	T	23. *So for* each pair, *why does one come from the* father *and one come from the* mother?	F (expand)
172	T	*Here, how many pairs are there? How many pairs of* **homologous chromosomes** *are there*?	I (open Q)
173	S1	Two.	R
174	T	Two pairs.	F (acknowledge)
176	T	**How many sets** *are there?*	I (open Q)
177	S1	*Two.*	R
178	T	*Two. Our cells have two sets?* **How many sets do sperms and eggs have?**	F (expand)
179	S1	*One set.*	R'
183	T	**Diploid cells, haploid cells.** *How many* **haploid cells** *are there in humans*?	I (open Q)
184	S1	*Only two kinds.*	R
185	T	**Which kind of** *haploid cells?*	F (expand)
186	S1	Sperm and Egg.	R'
187	T	*Other cells, skin, bones, hair, are all diploid cells.*	F' (ignore)
188	T	**Understand what I mean?**	I (close Q)
189	T	**Why is it called** diploid cells?	I (open Q)
190	T	*Because when you find a* chromosome, *it always has a partner. Always in a pair. They are called* diploid cells.	I (statement)
195	T	*Okay,* **For the same size** chromosomes *paired together, what are they called? [Cantonese]*	F'(expand)
198	S1	Homologous chromosomes. `	R'''
199	T	*That means the* chromosomes have the same shape and size, *or sometimes you say the same* structure and size.	F'''(ignore)

Excerpt 4C Students' language outputs in excerpt 4A

Turn	Speaker	Text	Grammatical categories	IRF sequence
165	S1	23.	Number	R
167	S1	23.	Number	R
173	S1	Two.	Number	R
177	S1	*Two.*	Noun phrase (Cantonese)	R
179	S1	*One set.*	Noun phrase (Cantonese)	R'
184	S1	*Only two kinds.*	Noun phrase (Cantonese)	R
186	S1	Sperm and Egg.	Nouns	R'
192	S1	Paternal chromosome.	Noun (technical)	R
194	S1	Maternal chromosome.	Noun (technical)	R'
196	S1	Homologous chromosomes	Noun (technical)	R''
198	S1	Homologous chromosomes.	Noun (technical)	R'''

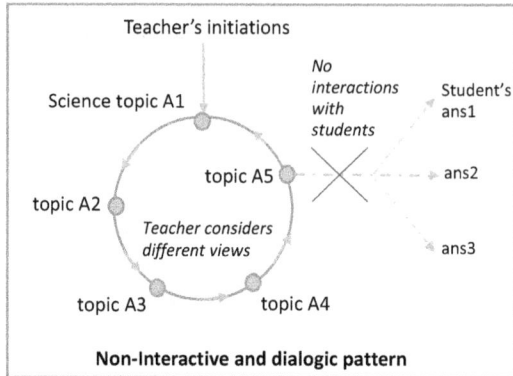

Figure 5.9 Conceptual diagram of the interactive and authoritative pattern no. 3.

So, we can see that after the students gave different answers in science topics A, B, and C, the teacher decided which was consistent with her planned science explanation and focused on it by asking more questions. Just like in the previous examples, the teacher only focused on students' answers if they aligned with her planned science explanation, leading to her asking more questions. This lesson again can be characterised as an interactive and authoritative science lesson (Mortimer & Scott, 2003; Figure 5.9).

5.2.5　*Summary: analysis of classroom interactions in the four biology lessons*

Figures 5.10 and 5.11 summarise the overall count of the IRF patterns in the four biology lessons in early–full and late–partial EMI schools and in Grades 10 and 11. To arrive at the figures, I calculated the overall distribution of the IRF pattern, the ratio between different forms of IRF (simple IRF and extended IRFRF), the percentage of the teacher's initiations, question types in the teacher's initiations (closed question/open question), the percentage of students' responses, and teacher feedback moves.

The data shows that there were more IRF patterns (22%) in the classroom interaction from the early–full EMI lessons than in the late–partial EMI lessons, with a higher ratio of extended IRF (45%) and simple IRF (55%). There was a relatively lower percentage of teacher's initiations (25%), with closed questions (4%) and open questions (16%), but slightly more students' responses (31%) and teacher's feedback (30%). In the late–partial EMI schools, there were fewer IRF patterns (12%), with a higher ratio of simple IRF (72%), and extended IRF (45%). There were significant numbers of teacher initiations (86%), with closed questions (7%) and open questions (24%), but slightly fewer students' responses (22%) and teacher's feedback (27%).

Biology lessons

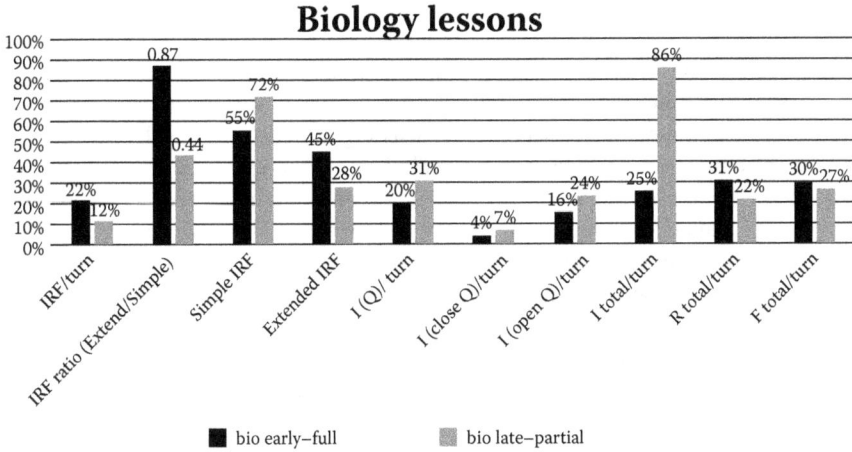

Figure 5.10 Distribution of IRF patterns between early–full and late–partial EMI schools.

Grade 10 vs Grade 11 in each group

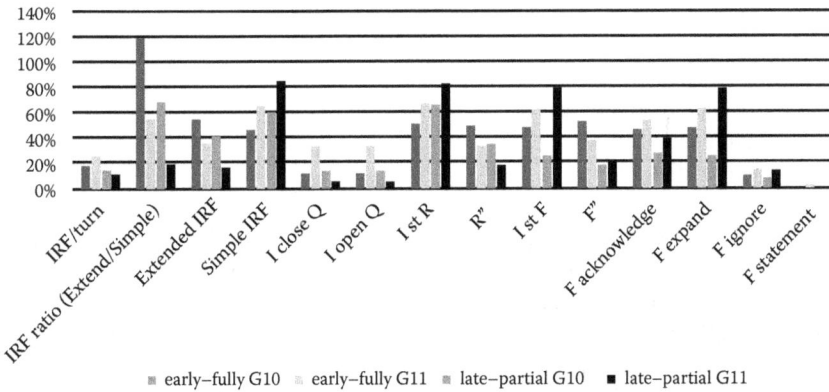

Figure 5.11 Distribution of IRF patterns between Grades 10 and 11 in each school group.

There were also similarities between Grades 10 and 11 in both types of schools in the four biology lessons. Regarding the similarities, there were more extended IRF sequences in Grade 10 and more simple IRF sequences in Grade 11. Additionally, there were more students' responses in Grade 11.

There were also differences between Grades 10 and 11, such that there were more IRF sequences in Grade 10, more closed questions and open questions in Grade 11 in teacher's initiations and more feedback in Grade 11 for the early–full EMI schools. However, an opposite pattern was found in the late–partial EMI schools. They showed fewer IRF sequences in Grade 10, fewer closed questions and open questions in Grade 11 in teacher's initiations, and less feedback in Grade 11.

In terms of the kind of language students produced in the classroom interactions, students from both early–full EMI and late–partial EMI schools predominately produced incomplete sentences consisting of short, technical, and abstract nouns or noun phrases referring to scientific items when responding to their teachers. In most of these exchanges, the teacher did not encourage students to produce a complete sentence or make any attempts to correct students' language mistakes or provide feedback to students about how to acquire the language of science in the L2 context. These patterns suggest that the science classroom across both types of schools fails to provide students with an opportunity to develop their English skills, or their science knowledge, through interaction with teachers.

5.2.6   The relationship between IRF patterns and the communicative approaches in science classrooms

Mortimer and Scott's (2003) four communicative approaches in science classrooms categorise science teaching and interaction along two continua: (1) from non-interactive to interactive and (2) from authoritative talk to dialogic talk. The endpoints of these continua provide four categories, i.e., four types of teaching interactions:

- non-interactive–authoritative
- non-interactive–dialogic
- interactive–authoritative
- interactive–dialogic.

Although this framework helps us categorise the nature of the interactional sequences in different EMI science classrooms, the approach does not provide a comprehensive account of the nature of interaction at the micro level. In particular, it does not explain the subtle differences in interactions, for example, whether the science teacher interacted with the students or not, or if the teacher considered the students' ideas as the lesson proceeded through interaction. Perhaps Mortimer and Scott's model only takes a synoptic view in categorising interaction as a finished product without providing further explanation of the dynamic unfolding in the interaction, whether the teacher takes advantage of opportunities in interaction to help the students learn science content and language.

To complement Mortimer and Scott's (2003) categories, I also analysed features of the dynamic unfolding in IRF interactions at the micro level, specifically to provide a comprehensive account of how the teacher interacted with the students to try to make the scientific ideas more accessible to students and to provide students with opportunities to learn the content and language. Based on the qualitative analysis carried out up to this point, Table 5.2 lists the features in terms of IRF patterns from the four biology EMI science classrooms. This analysis supplements Mortimer and Scott's (2003) framework. Table 5.3 summarises the relationship between IRF patterns and Mortimer and Scott's categories, whereas Tables 5.4 and 5.5 provide a comparison of the interactional features of the lessons that had simpler and more extended IRF sequences in the early–full and late–partial EMI biology lessons.

Table 5.2 (1) The observed interaction features for non-interactive and authoritative lessons in supplement to Mortimer and Scott's framework

A. Non-interactive and authoritative

Mortimer and Scott's definition	In a non-interactive and authoritative lesson, the teacher only presented a specific view of the science phenomena without any interactions with students.
The IRF analysis supplement to Mortimer and Scott	• long period of teacher's initiations with lots of statements; • less or nearly no interactions, absence of IRF, particularly extended IRF; • only a few comprehension checks, realised by closed Qs; • little to no students' responses; • no teacher's feedback built on students' answers; • science pedagogy where the teacher simply introduced his own way of understanding scientific phenomena, without considering students' ideas, with no exploration of the meaning, and language of science in the interaction; • the teacher did not unpack or repack the abstract scientific knowledge; • the teacher insisted on transferring his way of interpretation of the science topic; • no opportunity for L2 learning.

Table 5.2 (2) The observed interaction features for non-interactive and dialogic lessons in supplement to Mortimer and Scott's framework

B. Non-interactive and dialogic

Mortimer and Scott's definition	In non-interactive and dialogic lessons, the teacher considers different points of view, guiding students to explore, and work on different views of a science phenomenon without any turn-taking interactions with the students.
The IRF analysis supplement to Mortimer and Scott	• long period of teacher's initiations, with lots of statements; • a limited amount of interaction between teachers and students, only in the form of IRF, particularly more simple IRF; • many teachers' initiations, realised by closed Qs, or statements to only inform the teacher's voice or sometimes to promote students' short answers, mainly for comprehension checking;

(Continued)

Table 5.2 (2) (Continued)

B. Non-interactive and dialogic
• limited students' responses, followed by teacher's feedback to acknowledge students' answers in the direction of the teacher's planned explanation; • because students' answers are rarely considered, none of the students' ideas were explored or followed up; • the teacher only directed students towards his or her planned idea, yet encouraged students to consider different points of view, but without any turn-taking interaction with the students; • limited exploration of meaning and language of science in the interaction; • the teacher only unpacked the abstract scientific knowledge but also worked on different views of a scientific phenomenon; • the teacher asked for students' ideas but transferred his way of interpretation of the science topic, without much engagement by the students in their own science interpretation; • some opportunities for understanding content learning, but not much opportunity to practise the language.

Table 5.2 (3) The observed interaction features for interactive and authoritative lessons in supplement to Mortimer and Scott's framework

C. Interactive and authoritative	
Mortimer and Scott's definition	In an interactive and authoritative lesson, the teacher leads students through a great deal of turn-taking through a sequence of questions and answers to achieve one specific view on a scientific phenomenon.
The IRF analysis supplement to Mortimer and Scott	• long sequence of short periods, with lots of teacher's questions and statements; • but a considerably large amount of interaction between teacher and students, in the form of IRF, particularly more simple IRF; • short students' answers; • students' ideas were acknowledged by the teacher, followed up, with the teacher's feedback; • many teacher's initiations, realised by closed Qs, or statements to promote students' short answers, mainly for comprehension checking, rephrasing what students could not understand (perhaps due to students' limited English ability) and topic shifting to explore students' ideas;

(Continued)

Table 5.2 (3) (Continued)

C. Interactive and authoritative
• a great deal of students' responses, followed by teacher's feedback to acknowledge students' answers in the direction of the teacher's planned explanation; • views inconsistent with the teacher's planned ideas were further explored and discussed; • the teacher discussed with students what had been explored in the lesson, with the exploration of meaning and language of science in the interaction; • the teacher only unpacked the abstract scientific knowledge; • lots of opportunities for understanding content learning, but not much opportunity to practise the language.

Table 5.2 (4) The observed interaction features for interactive and dialogic lessons in supplement to Mortimer and Scott's framework

Interactive and dialogic	
Mortimer and Scott's definition	In an interactive and dialogic lesson, the science teacher listens to and takes account of his or her students' points of view. Then, the teacher explores different views with students to generate new meanings, ask real questions, and work on different points of view on a scientific phenomenon.
The IRF analysis supplement to Mortimer and Scott	• long sequence of short periods, with lots of teacher's questions and feedback; • considerably large amount of interactions between teacher and students, in the form of IRF, particularly more extended IRF; • extended students' answers; • students' ideas were acknowledged by the teacher, followed up, with the teacher's feedback; • a great deal of teacher's initiations, realised by open Qs, to promote students' long answers, mainly for exploration of meaning and practising the language (perhaps due to students' adequate English ability); • a great deal of students' responses, followed by teacher's feedback to acknowledge students' answers; • students' ideas inconsistent with the teacher's planned ideas were further explored and discussed; • teacher and students constructed knowledge; • the teacher unpacked and repacked the abstract scientific knowledge with the students; • the teacher discussed what had been explored in the lesson with the students, with the exploration of meaning and language of science in the interaction. • à lots of opportunities for understanding content learning, and many opportunities to practise the language

Table 5.3 The relationship between IRF patterns and the four communicative approaches in Mortimer and Scott's (2003) framework

Non-interactive and dialogic	*Interactive and dialogic*
Teacher presents only one specific view of a scientific phenomenon without any interaction with the students	*Teacher leads students through a great deal of interaction, by a sequence of questions and answers to achieve one specific view on a scientific phenomenon*

Non-Interactive and authoritative pattern

Interactive and Authoritative Pattern

– Long period of teacher's turns	– Long period of teacher's turns;
– Little or complete absence of IRF	– A considerable amount of IRF (more simple IRF);
– No students' responses	– Many teacher's initiations, realised by closed Qs to promote student's short answers;
– Mainly comprehension checks, by closed Qs	– Students' responses were acknowledged if in line with the teacher's views in the teacher's feedback;
– No teachers' feedback	– Ideas inconsistent with the teacher's planned ideas were NOT explored and discussed.

Example of lesson (Chemistry):

T: Last time we stopped at the Fractional Distillation of Petroleum [Topic 1]…

S: …

T: Today we will start at part 4. The users of the petroleum fractions [Topic 2]…

S: …

T: Hydrocarbon is a compound made up of hydrogen and carbon atoms [Topic 3] …

S: …

T: For incomplete reactions, the oxygen supply is poor. For the products, water produced, for the incomplete reaction, water is also produced [Topic 4]…

S: …

Example of lesson (Chemistry):

T: *Do you remember the two types of isomers we have learnt?*

S: Structural isomers and stereoisomers.

T: Right, structural isomers *have 3 sub-types, do you remember?*

S: *What is the daily implication? [ignored by teacher]*

T: *What is the* definition *of* isomers?

T: *Just to remind you, their* molecular formulas *are the same.* Molecular formulas *are the same.*

S: *3 carbons, 5 H 1 O. In fact, 5* isomers *have the same* molecular formulas.

(Continued)

Table 5.3 (Continued)

Non-interactive and dialogic	Interactive and dialogic
Teacher considers different points of view, guiding students to explore, and work on different views of a science phenomenon without any turn-taking interactions with the students	*Teacher listens to, and takes account of the students' points of view. Together with the students, he/she explores different views, generates new meanings, asks real questions, and works on different points of view on a scientific phenomenon.*

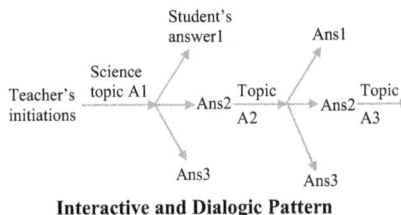

Non-Interactive and Dialogic Pattern

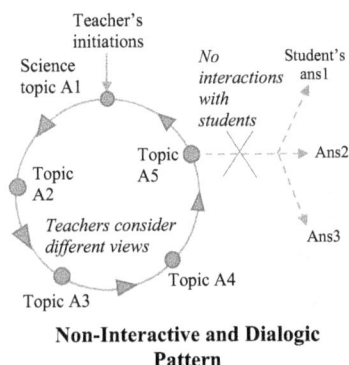

Interactive and Dialogic Pattern

– Long sequence of short periods of teacher's turn;	– Long sequence of short periods, with lots of teacher's questions and feedback;
– Minimal IRF;	– A considerably large amount of IRF (more extended IRF);
– Short students' answers;	– Extended students' answers;
– Students' ideas were not acknowledged by the teacher, nor followed up, with the teacher's feedback;	– Students' ideas were acknowledged by the teacher, and followed up, with the teacher's feedback;
– Many teacher's initiations;	– A great deal of teacher's initiations, realised by open Qs, to promote students' long answers, mainly for exploration of meaning and practising the language;
– Realised by closed Qs, or statements mainly for comprehension checks	– Ideas inconsistent with the teacher's planned ideas were further explored and discussed

Example of lesson (Chemistry):

T: Today we will look at Haloalkane first. From our syllabus, we have only one reaction for our haloalkane, in between alcohol. We look at our book. It is on page 12.

Example of lesson (Chemistry):

T: Last time we are talk about the four reactions of alkalis, four chemical reaction, do you still remember which four chemical reaction are there?

S: Neutralisation.

(*Continued*)

Table 5.3 (Continued)

Non-interactive and dialogic	Interactive and dialogic
S: Okay.	T: Neutralisation, very good. Neutralisation. So what is being reacted for neutralisation? Alkalis react with what? Alkalis react with what we will call it neutralisation?
T: This is the haloalkane. Haloalkane is, there is a halogen atom bonded to the carbon. And then, we look at this structure first.	
S: Yes.	S: Alkali react with acid
T: For halogen atom, which is electronegative, it tends to draw the bonded electrons towards itself. So X, Halogen atom, carry partial negative.	T: Yes, good. Alkali react with acid, so this is called neutralisation. And then for the second one, what is the second reaction we have learn?
S: Yes.	S: Non-metal oxide
T: And then the carbon become electron deficient.	T: Non-metal oxide, what is the non-metal oxide we talk about?
	S: Carbon dioxide
	T: Carbon dioxide, yes CO_2, reaction with CO_2. What is the main product that will be given out? What is the main product?
	S: Carbonate ion
	T: Yes. It will form some carbonate ion. So this is the carbonate ion: Especially for the calcium carbonate, it is insoluble in water, so remember this one.

Table 5.4 Comparison of the observed interaction features between a lesson with more extended IRF in early–full and late–partial EMI classrooms

Example	More extended IRF in early–full EMI biology	More extended IRF in late–partial EMI biology
Observed features	Short period but long sequence More closed Qs, teacher's feedback Students' input, for less in-depth exploration of meaning and more practising of language	Long period, but short sequence More open Qs, teacher's feedback The role of L1 in interactive classrooms à higher Qs, translation, daily examples, students input, for more in-depth meaning exploration and not much practising of language

Table 5.5 Comparison of the observed interaction features between a lesson with simpler IRF in early–full and late–partial EMI classroom

Example	*Simple IRF in early–full EMI biology*	*Simple IRF in late–partial EMI biology*
Observed features	Long period, short sequence More closed Qs, check for comprehension Teacher's acknowledgement mainly in feedback	Long period, short sequence More closed Qs, checks for comprehension Teacher's acknowledgement mainly in feedback L1, more translation for ensuring understanding of technical terms

5.3 Analysing the language of science in EMI science classrooms

Here, I analyse the language of science used between teachers and students during classroom interactions. The language of science comprises the language choices that the teacher and students use during interactions to develop their scientific knowledge. Using the integrated framework from Kong and Hoare (2012) and Fang (2006), explained in Chapter 3, these languages of science features are classified according to grammatical categories.

5.3.1 Early–full EMI biology lesson

In this lesson, students learnt about the basic concept of antibiotics and the historical development of penicillin (Alexander Fleming). The teacher drew on students' daily experiences with antibiotics and related science concepts (bacteria, pathogens, penicillin) to explain the treatment and biological mechanism for protecting humans from bacterial infection. At the end of the lesson, the teacher reminded students of the correct use of antibiotics. Table 5.6 summarises the language of school science used in the teacher's talk.

5.3.2 Late–partial EMI biology lesson

This lesson was about the immune system, where students learnt the basic concept of the mechanism of the human immune system. In this lesson, the teacher introduced the topic by asking students to consider related concepts (non-infectious disease, pathogens, infectious disease) and later asked the students to distinguish the differences in terms of the biological meanings of these terms. Then, the teacher asked the students to draw on their everyday experiences and explain the mechanism of the human immune system by relating it to the concepts learnt. Table 5.7 summarises the language of school science used in the teacher's talk.

Table 5.6 The language of science: subject-specific vocabulary

Characteristics of academic language	*Specific features of the language in school science*	*The language of science*
a. Subject-specific vocabulary	1. Technical vocabulary	Bacteria; virus; pathogens; penicillin; physiology; agar plate; petri dish; pipette; spreader; fungi (sg. fungus); yeast; cell division; genetic variation; nucleic acid; protein synthesis; respiration; H7N9; bacteriophage (a virus which parasitises a bacterium by infecting it and reproducing inside it); membrane; DNA (deoxyribonucleic acid); RNA (ribonucleic acid); host cells; influenza; Ebola; black death 黑死病 (the most devastating pandemic in human history); *Staphylococcus aureus* 金黄葡萄球菌 (a Gram-positive coccal bacterium that is a member of the Firmicutes, and is frequently found in the respiratory tract); Methicillin-resistant *Staphylococcus aureus*, MRSA 耐抗生素金黄葡萄球菌; Multimorphemic words with Latin or Greek origins: microorganisms = mirco+organ+ism+s antibiotics = anti+biotic+s; antigens = anti+gen; mitotic = mito+tic; antitoxin (an antibody that counteracts a toxin) = anti+toxi(c)+in; ribosome = ribo+some
	2. Ordinary words with non-vernacular meaning and usages	temperature n. (body temperature; the degree or intensity of heat present in a substance) incubate v. (keep bacteria at a suitable temperature so that they develop) (developing an infectious disease before symptoms appears) bacterial colonies n. (a community of animals or plants of one kind living close together) (country under full political control) words with multiple grammatical functions: culture n. (cultivation of bacteria; plants); culture v. (maintain cells in conditions suitable for growth)
	3. Words with multiple grammatical functions/ grammatical words with multiple usages	'or': a) denotes choices or alternatives: e.g. we will incubate them in an incubator for a certain time like 24 hours **or** 48 hours at a certain temperature b) introduce words/phrases that define, paraphrase, or synthesise a preceding term, usually preceded by a comma, e.g. for example 0.1 mL, using dropper, **or** we call it pipette for advanced tools

(*Continued*)

Table 5.6 (Continued)

Characteristics of academic language	Specific features of the language in school science	The language of science
b. Complex noun phrases	4. Ellipsis	T: No other bacteria contaminate this agar plate. Then how do we culture it? We can see <u>this process</u> *(culture of bacteria on the agar plate)*
	5. Subordinate clauses (non-finite clauses, elided subordinate clauses)	When *no cell wall is formed*, no reproduction occurs in bacteria. When *antibiotic use increases*, the resistance form and the percentage of resistant form will increase. They have some different antibiotics and placed on the agar plate which is spread by a certain kind of bacteria, which means before that, *before adding these antibiotics*, some bacteria are spread here.
	6. With-prepositional phrases	Later on if we continue to reproduce, *with the same genetic materials*, it continues to reproduce *with the same genetic materials.*
c. Nominalisations and grammatical metaphors	7. Abstract noun	Infection (infect); contamination (contaminate); incubation (incubate); dilution (dilute); DNA replication (repeat); diffusion (diffuse); reproduction (reproduce); mutation (mutate); formation (form);
d. Language of knowledge relationship	11. Passive voice	Second, how do antibiotics kill bacteria? Only bacteria, okay, *can be killed by antibiotics*, but antibiotics do not kill viruses.
e. Lexical density	71 content words /393 total words × 100 = 18%	

Table 5.7 The language of science

The characteristics of academic language	Specific features of the language in school science	The language of science
a. Subject-specific vocabulary	Technical vocabulary	non-infectious, infectious, pathogens, cardiovascular diseases, cardiac muscle, vascular bundle, coronary heart disease, bypass surgery, cholesterol, plaque, angioplasty, sternum (breastbone), vein, bacterial infection <u>words with Latin or Greek origin:</u> vascular (L); coronary (L); sternum (G);

(*Continued*)

Table 5.7 (Continued)

The characteristics of academic language	Specific features of the language in school science	The language of science
	2. Ordinary words with non-vernacular meaning and usages	Tube n.; transportation; object - a long hollow cylinder; plaque n.; an ornamental tablet; small raised patch resulting from local damage; Vein n; any of the tubes forming part of the blood circulation system; a fracture in rock containing a deposit of minerals; distinctive quality, style, tendency in singing <u>words with multiple grammatical functions:</u> clot n; a thick mass of coagulated liquid; clot v; cause to form clots/cover (something) with sticky matter
b. Complex noun phrases	4. Ellipsis	T: … so <u>this definition (</u>cardiovascular disease is not completely related to the heart) telling you that the disease is possibly related to the heart or blood vessels
c. Nominalisations and grammatical metaphors	7. Abstract noun (verbs à nouns) (product: grammatical metaphor; process: nominalisation)	Infection n. (infect v)
d. Language of knowledge relationship	11. Passive voice	Can you tell me what sort of risk can <u>be</u> found during this kind of surgery?
e. Lexical density	157 content words /509 total words × 100 = 31%	

5.4 Summary

This chapter has described the analytical framework I used to analyse the nature of the EMI science classroom in Hong Kong. Drawing on a number of analytical frameworks relating to science education and applied linguistics, I explored the nature of classroom interaction and used authentic accounts of EMI science classrooms as examples of Mortimer and Scott's four communicative categories (e.g., interactive/non-interactive and dialogic/authoritative). The four biology lessons from the early–full and late–partial EMI schools showed some differences in the nature of the content covered in the four lessons. These four lessons used interactive communication to gather students' ideas, but the teachers tended to be non-interactive and authoritative in choosing which students' responses to address, follow up on, or ignore. In Mortimer and Scott's framework, these four lessons were generally in the dialogue-authority dimension, indicating which of the students' science ideas these science teachers chose to develop. Most of the observed sequences between teachers and students took the form of simple IRF exchanges.

Both early–full EMI and late–partial EMI students used mainly short, technical, and abstract nouns or noun phrases to express some incomplete sentences. However, teachers rarely encouraged students to produce complete sentences, nor did they attempt to correct students' language mistakes or provide feedback to students about their L2 language production.

Observing the characteristics of the language in teacher and student talk, the language of science observed in the four biology EMI science classrooms can be categorised as 1) subject-specific vocabulary, 2) nominalisations and grammatical metaphors, 3) complex noun phrases, 4) language of knowledge relationships and 5) high lexical density. The measurement of lexical density serves as an indicator for analysing the level of complexity of teacher and student talk in EMI science classrooms. I also illustrate this with three examples of biology teachers asking higher-order questions, explaining words with similar meanings and using synonyms to describe a biological concept.

In addition, two IRF analyses illustrated how science teachers in Hong Kong developed students' thematic understanding of science topics through IRF interactions. Teachers intervene with students' responses at the feedback stage by either expanding, acknowledging, or ignoring students' ideas in developing students' scientific stories. Teachers' modified input in their talk during the IRF exchanges serves as an important way of providing extended comprehensive feedback to students in developing their science knowledge.

In this study, classroom interactions in EMI science classrooms were authoritative, often with the teacher controlling the interactions to achieve his or her particular view of the science topic. There was less teacher–student interaction, with some simple IRF patterns. Students rarely initiated interactions, mainly answering questions at the teacher's invitation to show their understanding of the science topic.

In Chapter 6, I will present an interview-based survey of teachers' perceptions of their EMI experiences. The interview data will reveal teachers' perceptions of teaching science in EMI classrooms, the linguistic challenges they may have encountered, and their coping strategies when teaching in English.

Note

1 Each of the lesson transcripts was analysed and coded into Initiation-Response-Feedback (IRF) sequences. Each IRF sequence can be categorised as a simple IRF or extended IRF pattern. The extended IRF pattern consists of a simple IRF pattern prolonged by more responses from students and subsequent teacher feedback. The ratio of IRF is the total number of IRF patterns divided by the total number of extended IRF patterns.

References

Fang, Z. (2006). The language demands of science reading in middle school. *International Journal of Science Education, 28*(5), 491–520.

Kong, S., & Hoare, P. (2012). The development of academic language proficiency: Challenges for middle school immersion in Hong Kong and Xi'an. *International Education, 41*(2), 88–109.

Lemke, J. L. (1990). *Talking science: language, learning and values*. Norwood, NJ: Ablex Publishing Corporation.

Mortimer, E., & Scott, P. (2003). *Meaning making in secondary science classrooms*. Berkshire, England: Open University Press.

6 Science teachers' perceptions of their EMI experiences

6.1 Questionnaire with science teachers on their perceptions of EMI teaching

This section presents a detailed analysis of the questionnaire data and findings on the teachers' perceptions of their English medium instruction (EMI) teaching in science classrooms.

6.1.1 Demographic characteristics of the teacher respondents

A total of 19 teachers returned their questionnaires, resulting in a response rate of 63%. The questionnaire consists of six parts: demographic characteristics; the choice of language of instruction in different science teaching activities; the most important and demanding English language skills students needed to study science in English medium; reflections on science teaching practices in the EMI context; the language challenges students faced in learning science in English; and the usefulness and frequency of the teachers' coping strategies. For demographic characteristics, the teacher respondents were asked to report their age range, gender, science subject responsibilities, number of years of total teaching experience, different curricula taught (e.g., HKDSE, A-Level, CE), self-reported English language skills and whether they had received any EMI training.

6.1.2 Questionnaire results (correlation analyses)

Prior to conducting correlation analyses, preliminary analyses were performed to check the assumptions of normality, linearity and homoscedasticity. Since the assumptions were violated, the relationship between teachers' background characteristics and different variables of teachers' responses about their perceptions of EMI teaching (as indicated by their average total scores for each survey item on a five-point scale) was investigated using Spearman's rho coefficient.

DOI: 10.4324/9781003001454-7

6.1.2.1 *Correlations between teachers' demographic characteristics and their perceptions of EMI teaching*

A Spearman's rank-order correlation was run to determine the relationship between the 19 teachers' demographic characteristics and their perceptions of EMI teaching. The results are shown in Table 6.1.

The results with statistical significance are as follows:

- When it comes to language choice in activities, teachers' self-reported overall English scores ($r = -0.574$, $p = 0.01$) and their English-speaking skills ($r = -0.61$, $p = 0.006$) were strongly negatively correlated. This suggests that teachers with lower self-reported English scores, particularly those with poorer speaking skills, tend to use more Cantonese in science teaching activities.
- When it comes to teachers' views on the importance of English, it was found that the MOI of the schools that teachers were teaching in was negatively correlated with their views towards the importance of English language skills ($r = -0.663$, $p = 0.002$). This possibly suggests that teachers in late–partial EMI schools tend to rate the importance of English language skills lower than teachers in early–full EMI schools.
- On teachers' language awareness, a number of teachers' characteristics were strongly correlated. These were school MOI ($r = -0.591$, $p = 0.008$), teachers' overall English scores ($r = 0.568$, $p = 0.011$), as well as their individual language skills including listening ($r = 0.487$, $p = 0.035$), speaking ($r = 0.599$, $p = 0.007$), reading ($r = 0.619$, $p = 0.005$) and writing ($r = 0.522$, $p = 0.022$). This perhaps suggests that teachers in early–full EMI schools, with higher self-reported English scores as well as superior individual language skills, tend to have a higher level of language awareness of teaching science through English than those in late–partial EMI schools.
- To summarise, teachers with lower self-reported English scores tend to use more Cantonese in science teaching activities. Teachers with EMI training, but less teaching experience at senior levels, have a higher level of teaching belief in EMI. Those teachers with lower self-reported writing scores in English tended to perceive the coping strategies as useful. Early–full EMI teachers, with higher self-reported scores for overall language proficiency as well as the four individual language skills, and more language awareness of teaching science through English, have the tendency to rate the importance of English language skills higher, as opposed to late–partial EMI teachers.

6.1.2.2 *Correlations among the eight questionnaire scales*

A Spearman's rank-order correlation was run to determine the relationship among the eight scales. The results of the correlations among the eight questionnaire scales are shown in Table 6.2. The results with statistical significance are as follows:

- Teachers' language awareness was negatively correlated with their choice of instructional language in different science teaching activities ($r = -0.587$, $p = 0.008$).

Table 6.1 The correlations between teachers' demographic characteristics and their perceptions of EMI teaching (*n* = 19)

	1.Lang. choice in science activities	2.English demanding	3.English important	4.Language awareness	5.Belief	6.Challenges	7.Freq. strategy	8.Useful. strategy
a. School Type	n.s.	n.s.	-0.663**	-0.591**	n.s.	n.s.	n.s.	n.s.
b. Sex	n.s.	n.s.	n.s.	n.s.	n.s.	n.s.	n.s.	n.s.
c. Age	n.s.	n.s.	n.s.	n.s.	n.s.	n.s.	n.s.	n.s.
d. University degree	n.s.	n.s.	n.s.	n.s.	n.s.	n.s.	n.s.	n.s.
e. EMI training	n.s.	n.s.	n.s.	n.s.	0.462*	n.s.	n.s.	n.s.
f. English scores	-0.574*	n.s.	n.s.	0.568*	n.s.	n.s.	n.s.	n.s.
g. Listening	n.s.	n.s.	n.s.	.487*	n.s.	n.s.	n.s.	n.s.
h. Speaking	-0.610*	n.s.	n.s.	0.599**	n.s.	n.s.	n.s.	n.s.
i. Reading	n.s.	n.s.	n.s.	0.619**	n.s.	n.s.	n.s.	n.s.
j. Writing	n.s.	n.s.	n.s.	0.522*	n.s.	n.s.	n.s.	-.526*
k. Teaching experience (junior)	n.s.	n.s.	n.s.	n.s.	n.s.	n.s.	n.s.	n.s.
l. Teaching experience (senior)	n.s.	n.s.	n.s.	n.s.	-0.526*	n.s.	n.s.	n.s.
n. Teaching experience in EMI	n.s.	n.s.	n.s.	n.s.	n.s.	n.s.	n.s.	n.s.
m. Teaching experience in integrative science (junior)	n.s.	n.s.	n.s.	n.s.	n.s.	n.s.	n.s.	n.s.
l. Teaching science subjects	n.s.	n.s.	n.s.	n.s.	n.s.	n.s.	n.s.	n.s.
o. HKDSE	n.s.	n.s.	n.s.	n.s.	n.s.	n.s.	n.s.	n.s.
p. CE	n.s.	n.s.	n.s.	n.s.	n.s.	n.s.	n.s.	n.s.
q. A-level	n.s.	n.s.	n.s.	n.s.	n.s.	n.s.	n.s.	n.s.

Note: *$p < 0.05$; **$p < 0.01$; ***$p < 0.001$; n.s. = no statistical significance.

Table 6.2 Summary of the correlations among the eight questionnaire scales

	1	2	3	4	5	6	7	8
1. Lang. choice in science activities	n.s.	n.s.	n.s.	−0.587**	n.s.	n.s.	n.s.	n.s.
2. English demanding	n.s.		n.s.	n.s.	n.s.	n.s.	n.s.	n.s.	n.s.
3. English important	n.s.		n.s.	n.s.	n.s.	−0.466*	n.s.	n.s.	n.s.
4. Language awareness	−0.587**	n.s.	n.s.		n.s.	n.s.	n.s.	n.s.	n.s.
5. Belief	n.s.		n.s.	−0.466*	n.s.		n.s.	n.s.	n.s.	n.s.
6. Challenges	n.s.		n.s.	n.s.	n.s.	n.s.		n.s.	n.s.	n.s.
7. Freq. strategy	n.s.		n.s.	n.s.	n.s.	n.s.	n.s.	n.s.	0.756***
8. Useful. Strategy	n.s.		n.s.	n.s.	n.s.	n.s.	n.s.	n.s.	n.s.

Note: $*p < 0.05$; $**p < 0.01$; $***p < 0.001$; n.s. = no statistical significance.

This perhaps suggests that teachers who reported a higher level of language awareness tend to use more English in science teaching activities.

- Teachers' views on the importance of English language skills were negatively
  correlated with their teaching belief in EMI teaching ($r = -0.466$, $p = 0.044$).
  This possibly suggests that teachers who rated the importance of English language skills lower are likely to have a higher level of teaching belief in EMI.
- Teachers' views on the frequency of coping strategies were found to be positively
  correlated with their views on the usefulness of coping strategies ($r = 0.756$,
  $p < 0.0001$). This perhaps suggests that teachers who reported using coping
  strategies more frequently tend to find these strategies useful.
- The scales of teachers' language awareness investigated whether teachers make
  necessary steps in their teaching to make science more accessible in English using statements such as *I use special teaching methods to help my students understand the English vocabulary I use in my science teaching* and *I use a number of
  teaching strategies in teaching science through English that I would not use if
  I taught in Chinese.* In sum, teachers with a higher level of language awareness
  (i.e., teachers reported putting more effort in making science more accessible
  to students in English) tend to use more English in science teaching activities.
  Those with a higher level of teaching belief in EMI tend to rate the importance
  of English language skills lower. Those who used more frequent coping strategies are likely to find these strategies useful.

6.1.3   Questionnaire results (descriptive statistical analyses)

Due to the small number of participants ($n = 19$), I could only perform descriptive
statistical analyses on the teachers' responses. The major results of the teacher's
questionnaire are presented in Tables 6.3–6.8. These tables present the means and
standard deviations of each of the survey items. In addition, I performed a correlation analysis using the teachers' responses in both groups (early–full and late–
partial) on the eight questionnaire scales (as indicated by their average total scores
for each survey item on a five-point scale).

6.1.3.1 Questionnaire results on the eight aspects of teachers' EMI teaching

For the choice of instructional language in different science teaching activities, teachers in late–partial EMI schools reported using more Cantonese or code-switching between English and Cantonese in their EMI teaching. This is revealed in Table 6.3, which shows that the mean scores of teacher responses in late–partial EMI schools were higher than those of teacher responses in early–full EMI schools. However, the overall sum of scores for the 12 items about the choice of instructional language in science classrooms did not differ between the early–full and late–partial EMI schools. Table 6.3 summarises the teachers' responses to the choice of instructional language in different science teaching activities.

Table 6.3 The teachers' responses to the choice of instructional language in different science teaching activities

	Mean (S.D.)		
	Early–full	*Late–partial*	*Overall*
1. Teaching the whole class	1.83 (0.75)	3 (1.29)	2.63 (1.26)
2. Giving instructions to the students	2 (1.01)	3.38 (1.12)	2.95 (1.27)
3. Answering students' questions in front of the whole class	1.83 (0.753)	3 (1.29)	2.6 (1.26)
4. Discussing ideas with the whole class	1.83 (0.75)	3.08 (1.19)	2.68 (1.2)
5. Managing the classroom	2.5 (1.05)	3.54 (1.33)	3.21 (1.32)
6. Talking to individual students about their work	2.83 (0.75)	4 (0.82)	3.63 (0.96)
7. Taking part in teacher–student interactions	2.67 (0.52)	3.38 (1.39)	3.16 (1.21)
8. Explaining the definition of scientific vocabulary	2.67 (0.82)	3 (1.16)	2.89 (1.05)
9. Explaining a science phenomenon	2.5 (0.84)	2.92 (1.32)	2.79 (1.18)
10. Explaining a science phenomenon using everyday experience	2.67 (0.82)	3.08 (1.19)	2.95 (1.08)
11. Giving instructions to students for a lab experiment	2.33 (0.52)	3.08 (1.26)	2.84 (1.12)
12. Clarifying students' science misconception in a lab experiment	2.83 (0.75)	3.23 (1.09)	3.11 (0.99)
Overall score of instructional language in current science classrooms	2.38 (0.6)	3.22 (1.09)	2.96 (1.025)

Note: 1. The choice of instructional language was rated on a 5-point scale, where 1 = (always English), 2 = (usually English), 3 = (code-switching between English and Cantonese), 4 = (usually Cantonese), and 5 = (always Cantonese).

On the question of teachers' language awareness in teaching science through English, the teachers from the late–partial EMI schools demonstrated a relatively lower level of language awareness than those from the early–full EMI schools. This is illustrated in Table 6.4, wherein the mean scores of teachers' responses with respect to their language awareness in teaching science using English in the early–full EMI schools were higher than those in the late–partial EMI schools.

As shown in Table 6.5, the late–partial EMI teachers scored higher than the early–full EMI teachers in terms of their reflections of science teaching practices in the EMI context. However, the early–full EMI teachers had a higher ranking for Items 4 and 8 with respect to expectations of students' English levels. Although these teachers did not believe they have a role in teaching English, they felt strongly about their need to know how their students learn effectively through English. This is reflected in their responses to Item 4 ('Developing students' English proficiency is not an important part of a science teacher's role in an English medium school') and Item 8 ('Knowing how students can learn English effectively through science is useful to science teachers in an EMI school').

As for the language challenges found in learning science through English, the late–partial EMI teachers believed their students faced more language challenges. This is shown in Table 6.6, where the mean scores of the late–partial EMI teachers were higher than those of the early–full EMI teachers. However, the teachers from early–full EMI schools rated Items 6, 7, 14, 15, 16, 22, 24 and 26 higher. Table 6.6 summarises the teachers' responses to the language challenges for learning science in English.

Regarding the frequency of using coping strategies, the early–full EMI teachers reported that they used coping strategies more frequently. This is demonstrated in Table 6.7, where the mean scores of the teachers in early–full EMI schools were higher in their response to the frequency of using coping strategies than those in late–partial EMI schools. However, the teachers in late–partial EMI schools rated Items 1, 2, 8 and 5 higher.

With regard to the perceived usefulness of coping strategies, the early–full EMI teachers reported these strategies were more useful. This is revealed in Table 6.8, where the mean scores of the teachers in early–full EMI schools were higher in their responses to most of the items on the usefulness of coping strategies. However, the teachers in late–partial EMI schools rated Items 1, 5 and 8 higher. Table 6.8 summarises the teachers' responses to the usefulness of coping strategies.

6.1.4   Summary of the questionnaire analyses

The correlation analyses yielded the following findings. Firstly, the teachers with lower self-reported English scores have the tendency to use Cantonese in science teaching activities. Secondly, teachers with EMI training, but less teaching experience at senior levels, have a higher level of teaching belief. Those teachers with low self-reported writing scores in English tend to perceive the coping

Table 6.4 The teachers' responses to their language awareness in teaching science through English

	Mean (S.D.)		
	Early-full	*Late-partial*	*Both*
1. I use special teaching methods to help my students understand the English vocabulary I use in my science teaching.	3.83 (0.75)	3.83 (0.75)	3.37 (1.12)
2. I always pay a lot of attention to the teaching of vocabulary in my science lessons because it is conducted in English.	3.83 (0.75)	3.23 (1.01)	3.42 (0.96)
3. I regularly ask my students to read their English science textbooks on their own.	3.67 (1.21)	2.92 (1.32)	3.16 (1.30)
4. I regularly ask my students to read other science resources in English.	3.5 (1.05)	2.69 (1.18)	2.95 (1.18)
5. In my science teaching, I pay a lot of attention to help students developing their English while they learn science.	3.33 (1.03)	2.38 (0.77)	2.68 (0.95)
6. I use a number of teaching strategies in teaching science through English that I would not use if I taught in Chinese.	3.33 (1.03)	2.38 (1.12)	2.68 (1.16)
7. I find it helpful to use my knowledge of English grammar to support my students in learning science through English.	3.67 (0.82)	2.92 (0.95)	3.16 (0.96)
8. I find it helpful to use my knowledge of English vocabulary to support my students in learning science through English.	4.0 (0.63)	3.77 (0.93)	3.84 (0.83)
9. In my science lessons, I try to use English in ways that will help my students to learn English as they learn science.	3.67 (0.82)	2.85 (1.21)	3.11 (1.15)
10. In my science lessons, I regularly try to give my students opportunities to talk in English in order to help them learn both science and English.	3.67 (1.03)	2.77 (1.36)	3.05 (1.31)
11. I try to help my students use English when they speak to me (e.g., answering questions) in my science classes.	3.83 (0.75)	2.46 (1.45)	2.89 (1.41)
12. I regularly require my students to write long answers in English (one sentence or more) as part of their learning of science.	4.0 (1.27)	2.31 (1.18)	2.84 (1.43)
The overall score of teacher language awareness in teaching science through English.	3.69 (0.64)	2.82 (0.64)	3.10 (0.75)

Note: The teacher language awareness in teaching science was rated on a 5-point scale, where 1 = (untrue of me), 2 = (somewhat untrue of me), 3 = (neutral), 4 = (somewhat true of me), and 5 = (very true of me).

Table 6.5 The teachers' responses to their reflections of science teaching practices in the EMI context

	Mean (S.D.)		
	Early–full	Late–partial	Both
1. It is important for English medium science teachers to find ways of helping their students to deal with the English vocabulary they meet in science lessons.	1.83 (0.41)	3.0 (1.41)	2.63 (1.3)
2. English medium students need specialised help to understand the English vocabulary they used in science learning.	2.67 (0.52)	3.23 (0.73)	3.05 (0.71)
3. Reading the English science textbooks is an important way for students to learn the English they need for studying science.	2.17 (0.98)	2.85 (1.52)	2.63 (1.38)
4. Developing students' English proficiency is not an important part of a science teacher's role in an English medium school.	2.83 (0.75)	2.69 (0.95)	2.74 (0.87)
5. Helping students to understand English they use in science is one of the most important responsibilities for a science teacher in an English medium school.	2.33 (0.52)	3.08 (1.19)	2.84 (1.07)
6. All English medium science teachers need to know how to support their students in understanding the use of English in a science classroom.	2.17 (0.41)	2.38 (0.87)	2.32 (0.75)
7. It is important for science teachers to have some knowledge about the English language (e.g., grammar, phonology, semantic, genre) in order to support students' learning of science through English.	2.00 (0.63)	2.38 (0.87)	2.26 (0.81)
8. Knowing how students can learn English effectively through science is useful to science teachers in an EMI school.	2.5 (0.84)	2.38 (0.96)	2.42 (0.902)
9. Teachers who speak fluent English will motivate students to improve their uses of English in a science class.	2.17 (0.41)	2.23 (1.17)	2.21 (0.98)
10. Teachers who write good English will motivate students to improve their uses of English in a science class.	2.17 (0.41)	2.23 (1.09)	2.21 (0.92)
11. In English medium science lessons, students need regular opportunities to talk about science in English.	1.83 (0.41)	2.38 (0.87)	2.21 (0.79)
12. It is important for students to do learning tasks in science that require them to produce long spoken answers in English.	2.5 (0.55)	2.92 (0.76)	2.79 (0.71)
13. It is important for students to do learning tasks in science that require them to produce long written answers in English.	2.17 (0.41)	2.46 (0.66)	2.37 (0.60)
Overall score of the teachers' reflections on their science teaching practices in EMI context.	2.26 (0.19)	2.63 (0.63)	2.51 (0.56)

Note: The teachers' reflections on their science teaching practices were rated on a 5-point scale, where 1 = (untrue of me), 2 = (somewhat untrue of me), 3 = (neutral), 4 = (somewhat true of me), and 5 = (very true of me).

Table 6.6 The teachers' responses to the language challenges for learning science in English

	Mean (S.D.)		
	Early–full	Late–partial	Both
1. **Transmitting content knowledge through English**	2.83 (1.47)	2.85 (1.07)	2.84 (1.17)
2. **Teaching and explaining abstract scientific ideas in English**	3.5 (1.51)	4.15 (0.69)	3.95 (1.03)
3. **Teaching and explaining complex scientific ideas in English**	3.83 (1.47)	4.23 (0.60)	4.11 (0.937)
4. **Teaching students to distinguish vocabulary and technical terminology in English**	2.83 (1.33)	3.08 (0.76)	3.0 (0.943)
5. **Making students to answer to my questions in English**	3.33 (0.52)	3.46 (0.78)	3.42 (0.692)
6. **Making students to discuss themselves in English**	4.17 (0.75)	4.08 (1.12)	4.11 (0.994)
7. **Evaluating students' science understating in English**	3.17 (1.33)	3.08 (0.95)	3.11 (1.05)
8. **Transmitting scientific knowledge and skills to students in English**	2.67 (1.21)	3.15 (0.98)	3.0 (1.05)
9. **Providing opportunities for practice and giving corrective feedback for consolidation in English**	2.83 (1.17)	3.31 (0.86)	3.16 (0.96)
10. **Using questions to prompt students for higher-order cognitive thinking in English**	3.33 (1.21)	3.62 (0.87)	3.53 (0.96)
11. **Developing students' scientific skills for investigation study (testing hypothesis, designing working procedures, gathering data, performing calculation, drawing conclusion)**	3.0 (1.1)	3.62 (1.2)	3.42 (1.17)
12. **Encouraging students to generate new knowledge through group work by sharing their knowledge in English**	3.33 (1.03)	3.69 (0.76)	3.58 (1.12)
13. **Initiating students to apply existing knowledge to unfamiliar situations in English**	3.33 (1.03)	3.92 (0.76)	3.74 (0.87)
14. **Preparing teaching content for presentations in English**	2.17 (1.47)	2.08 (0.86)	2.11 (1.05)
15. **Dealing with different levels of English language proficiency/ skills of the students**	3.33 (1.21)	3.15 (1.07)	3.21 (1.08)
16. **Dealing with different levels of science achievements of the students**	3.5 (1.05)	3.0 (0.71)	3.16 (0.83)

(*Continued*)

Table 6.6 (Continued)

	Mean (S.D.)		
	Early–full	Late–partial	Both
17. **Dealing with diversity of existing science knowledge among students**	3.17 (1.72)	3.54 (0.66)	3.42 (1.07)
18. **Choosing appropriate language style for explaining a science concept**	2.5 (1.38)	3.0 (0.86)	2.84 (1.02)
19. **Choosing appropriate text organisation for explaining a science concept**	2.67 (1.37)	2.85 (0.801)	2.79 (0.98)
20. **Explaining topics after reading materials**	2.17 (1.17)	3.08 (0.76)	2.79 (0.98)
21. **Designing a test/an exam with consideration on the English demand for students (e.g., writing, M.C.)**	2.17 (1.602)	2.31 (0.48)	2.26 (0.93)
22. **Deciding the science topics and levels of difficulty in a test or an exam**	2.5 (1.64)	2.23 (0.44)	2.32 (0.95)
23. **Writing exam questions in English with consideration on students' English level**	2.17 (1.17)	2.31 (0.75)	2.26 (0.87)
24. **Assessing English language mistakes in written exams**	2.83 (1.17)	2.54 (0.88)	2.63 (0.96)
25. **Assessing students' understanding of the subject content**	2.17 (1.17)	2.54 (1.05)	2.42 (1.07)
26. **Making a balance between correcting students' mistakes in English and in science concepts**	2.83 (0.98)	2.69 (0.95)	2.74 (0.93)
Overall score of language challenges in teaching through English	2.94 (1.07)	3.13 (0.38)	3.07 (0.66)

Note: The language challenges for learning science in English were rated on a 5-point scale, where 1 = (strongly disagree), and 5 = (strongly agree).

strategies as useful. The early–full EMI teachers with higher self-reported scores in overall language proficiency and the four individual language skills are likely to prioritise the importance of English language skills. Further, they exhibit more language awareness of teaching science through English as opposed to the late–partial EMI teachers. Finally, the teachers with a higher level of language awareness tend to use more English in science teaching activities. Those with a higher level of teaching belief in EMI tend to rate lower the importance of English language skills and those who used coping strategies more frequently are likely to find these strategies useful.

Table 6.7 The teachers' responses to the frequency of using coping strategies

	Mean (S.D.)		
	Early–full	Late–partial	Both
1. **Providing Chinese translation of technical vocabulary**	2.67 (0.52)	3.46 (1.39)	3.21 (1.23)
2. **Providing Chinese translation of abstract English vocabulary (not technical, e.g., metaphor)**	3.0 (0.63)	3.38 (1.26)	3.26 (1.10)
3. **Providing more definitions and exemplification of technical words in English but not necessary to translate them into Chinese**	3.5 (1.05)	3.0 (1.08)	3.16 (1.07)
4. **Providing students with more English language instructions from their current English teachers**	3.17 (1.17)	1.85 (1.28)	2.26 (1.37)
5. **Providing students with additional English language course**	1.33 (0.52)	1.85 (1.41)	1.68 (1.20)
6. **Science teachers should receive more training on EMI teaching and the knowledge of English language**	3.0 (1.10)	2.54 (0.97)	2.68 (1.00)
7. **Providing additional English explanations and inputs from science teachers**	3.5 (1.05)	3.0 (0.91)	3.16 (0.96)
8. **Providing teaching notes to students in both Chinese and English**	1.5 (0.84)	1.77 (1.3)	1.68 (1.16)
9. **Providing students with more feedback and corrections on their English use in science**	3.5 (1.05)	3.31 (0.95)	3.37 (0.96)
10. **Collaborative teaching between English language teachers and science teachers on the same science topic**	2.5 (1.05)	1.8 (1.36)	2.0 (1.29)
11. **Students should attend tutorials and seek for study assistants outside school about their science and English**	3.17 (0.75)	2.38 (1.33)	2.63 (1.21)
Overall score of frequency of coping strategies rated by teachers	4.26 (3.57)	3.26 (2.50)	3.57 (2.81)

Note: The frequency of using coping strategies was rated on a 5-point scale, where 1 = (untrue of me), 2 = (somewhat untrue of me), 3 = (neutral), 4 = (somewhat true of me), and 5 = (very true of me).

The descriptive analyses of the questionnaire results for the eight aspects of teachers' experiences yielded the following findings. The late–partial EMI teachers reported that their students tend to use more Cantonese than English in science teaching activities. Moreover, they rated English language skills as more demanding than the teachers from the early–full EMI schools, but rated the same set of English skills lower in terms of importance than did the teachers from early–full EMI schools. This result reflects the same tendency as the students in late–partial schools (discussed further in Chapter 8). The late–partial EMI teachers also reported lower language awareness in teaching science through English,

Table 6.8 The teachers' responses to the usefulness of coping strategies

	Mean (S.D.)		
	Early–full	*Late–partial*	*Both*
1. Providing Chinese translation of technical vocabulary	3.17 (1.17)	3.23 (1.54)	3.21 (1.40)
2. Providing Chinese translation of abstract English vocabulary (not technical, e.g., metaphor)	3.67 (0.82)	3.46 (1.39)	3.53 (1.22)
3. Providing more definitions and exemplification of technical words in English but not necessary to translate them into Chinese	3.67 (1.03)	3.08 (1.04)	3.26 (1.05)
4. Providing students with more English language instructions from their current English teachers	3.17 (1.17)	2.38 (1.5)	2.63 (1.42)
5. Providing students with additional English language course	2.17 (1.6)	2.69 (1.7)	2.53 (1.65)
6. Science teachers should receive more training on EMI teaching and the knowledge of English language	19.5 (38.97)	3.0 (0.82)	3.11 (0.99)
7. Providing additional English explanations and inputs from science teachers	3.83 (1.33)	3.08 (0.86)	3.32 (1.06)
8. Providing teaching notes to students in both Chinese and English	2.33 (1.21)	9.69 (26.86)	2.26 (1.20)
9. Providing students with more feedback and corrections on their English use in science	3.67 (1.03)	3.38 (0.96)	3.47 (0.96)
10. Collaborative teaching between English language teachers and science teachers on the same science topic	3.5 (1.64)	2.23 (1.59)	2.63 (1.67)
11. Students should attend tutorials and seek for study assistants outside school about their science and English	3.5 (0.84)	2.69 (1.38)	2.95 (1.27)
Overall score of frequency of coping strategies rated by teachers	4.74 (3.82)	3.53 (2.64)	3.92 (3.01)

Note: The usefulness of coping strategies was rated on a 5-point scale, where 1 = (untrue of me), 2 = (somewhat untrue of me), 3 = (neutral), 4 = (somewhat true of me), and 5 = (very true of me).

suggesting that perhaps teachers in late–partial EMI schools tend to pay less attention to students' language issues.

However, the late–partial EMI teachers had a lower rating in their reflections about science teaching practices in the EMI context. This suggests that EMI had a negative impact on their teaching. This potentially explains why they rated most of the language challenges higher than the early–full EMI teachers, suggesting that the late–partial teachers believed their students encountered a number of language challenges when learning science through English. When asked about the frequency and the usefulness of coping strategies, the ratings from the teachers in late–partial EMI schools were lower than those in early–full EMI schools. This suggests that teachers in late–partial EMI schools tend to use coping strategies less frequently and find these coping strategies less useful.

6.2   Semi-structured interviews with science teachers

6.2.1   *Background of the teachers interviewed*

A total of 19 science teachers (6 physics teachers, 6 chemistry teachers and 7 biology teachers; 2 females and 17 males) participated in follow-up semi-structured interviews after lesson observations on the same teaching day. While six of them are from early–full EMI schools, the other 13 teachers are from late–partial EMI schools. All the participating teachers had obtained a science degree in their teaching area and held a postgraduate diploma in education (PGDE) from Hong Kong. One teacher, T9, had obtained a masters' degree in science education. All identified Cantonese as their first language. Their average teaching experience was 17.47 years (3–35 years). Yet, almost none of the participating teachers had any qualifications or had received any training in EMI, nor had they participated in any training related to teaching science through English. Only one teacher, T6, indicated in the questionnaire that she had attended a two-week professional development workshop about language across the curriculum at one of Hong Kong's universities. All the teachers had passed their English language subject before being admitted to university, which means that they possess English proficiency equivalent to a score of 6 or above in the International English Language Testing System examination. Each of the participating teachers was given a pseudonym and a code (T1–T19), which will be used in the subsequent analysis.

6.2.2   *Major themes from the teachers' semi-structured interviews*

Each interview consisted of three parts. First, the pre-observation interview aimed to explore the teaching purpose and learning objectives in the lessons that I was about to observe. Second, after the observations, teachers were asked to comment on the lessons. This part of the interview was guided by the following list of questions:

1   How did you feel about the lesson? Was the planned learning objective achieved? If yes, how? If not, what needs to be done?
2   What teaching and learning challenges did you experience?
3   Were there any specific events in the lesson that need to be clarified?
4   What has been the impact of EMI on classroom interactions?
5   What has been the impact of EMI on your teaching?

Third, in the second part of the post-observation interview, I asked the teachers for their views on EMI teaching, particularly their opinions on using English to teach senior secondary science, the language challenges they encountered, and the coping strategies that they used. The teachers were given an opportunity to express their views on other relevant topics such as how they perceive the new science curriculum in comparison with the previous seven-year curriculum. The results of the analysis of the teachers' interviews are presented in the following section based on four recurring themes.

6.2.3   *The results of the analysis of the teachers' interviews*

6.2.3.1   *Teachers' views on EMI teaching in science classrooms*

Late–partial EMI teachers strongly believe that the medium of instruction is merely a tool for transmitting knowledge rather than a means to prepare students to learn a second language. A late–partial EMI chemistry teacher noted that many science teachers are not good at English. Science teachers do not speak English solely in their classrooms.

Late–partial EMI teachers generally welcomed a mixed-mode instruction with Cantonese as the matrix language embedded with English scientific terms. The use of L1 in the science classroom allows teachers to explain logical relationships easily, helps them manage students with limited language ability, and also consolidates students' science knowledge as teachers can make the necessary connections with what students have learnt previously or what they see in their daily life. Mixed-mode instruction was cited as convenient for both teachers and students, especially weaker students. Mr. Yu, a chemistry teacher at a late–partial school, explained that:

> I generally agree with mixed-mode instruction in class. Cantonese helps consolidate students' knowledge and also it is more convenient for teachers; as you know, many science teachers are not very good at English. In fact, even for EMI schools, I don't think many teachers have completely used English language in class.
>
> (T9)

Mr. Chik, a physics teacher at a late–partial school similarly, noted that:

> I think language is a tool to achieve knowledge (in this case, which means Physics), so, it's okay to use any language as long as students understand … Although some students will take English papers, I think using Cantonese to explain physics concepts and the cause-and-effect relations is better for their understanding.
>
> (T10)

Mr. Chu, a chemistry teacher at a late–partial EMI school explained:

> I think the motivation or benefit [of using Cantonese] would be [teachers] can have more choices, because you can never force weaker students to adapt to 100% English.
>
> (T8)

In sum, the use of Cantonese is necessary for teachers that have to accommodate students with poor English proficiency, as well as consolidating students' science knowledge and getting them to think deeply in the late–partial EMI schools.

6.2.3.2 *Language challenges in teaching science through English*

Early–full EMI teachers stressed that English is the best possible language to teach scientific terms. They reported that it would be convenient to use English for naming the scientific terms, e.g., elements in the periodic table. For instance, Mr Lam, an early–full chemistry teacher, stated:

> If you ask me to tell the Periodic Table in Chinese, I can't… Except Chinese Language and Liberal Studies, all subjects are taught in English, so our ability to express and write in English is not a problem. However, if you ask me to express some concepts in Chinese, that will be hard.
>
> (T6)

Other teachers reported that delivering science content in English would help students recall what they learnt in other science subjects, which were also conducted in English.

However, when discussing the specific language challenges in different science subjects, teachers in both the early–full and late–partial EMI schools admitted that their students encountered a number of challenges. The EMI biology teachers from both groups expressed their concerns about students' analytical, reading, and communication skills. They highlighted the following language challenges:

- lack of analytical skills necessary to understand the examination questions, identify the important concepts or interpret abstract questions
- lack of motivation to read extensively about the learnt science concepts in textbooks or other sources like the internet
- lack of communication skills in presenting science ideas in a logical, coherent way with adequate use of scientific vocabulary and expressions
- excessive reliance on rote memorisation
- failure to express science ideas with correct English vocabulary.

Most of the early–full EMI biology teachers reported that their students did not have good analytical skills. The teachers found their students could not understand the examination questions, identify the important concepts or interpret abstract questions. When reading, students only skimmed through the bullet points and were rarely motivated to read the whole chapter of their science textbooks or to explore the relevant science content in other chapters.

In contrast to the early–full teachers, nearly all the late–partial EMI biology teachers believed their students could not communicate science ideas in English. They reported that although students could recall terms in Chinese, they generally did not have knowledge of English technical terms. Students were reluctant to speak English in the class and found it difficult to raise questions or interact with their peers in English. The teachers also expressed concerns about students not being able to describe and explain their science ideas in simple English. A late–partial EMI biology teacher pointed out that his students could not present any coherent science explanation in English with adequate vocabulary or scientific expressions. He reported that students knew the science ideas but failed to organise them in a logical and coherent manner. Students made inappropriate choices of vocabulary, often rendering their arguments incomplete. As Mr. Law, a late–partial biology teacher, explained:

> The biggest problem is students are bad at presenting these ideas in exams and tests. This problem is related to their English. They understand the ideas, but since they use inappropriate vocabulary or write incomplete answers, they can't get the marks. Sometimes when it comes to long questions, I can get their points, but their answers are just not coherent. They have so much information, but they don't know how to organise it into a logical argument or explanation.
>
> (T11)

Ms. Leung, a late–partial biology teacher, similarly noted:

> Students who understand the science concepts often can't express their ideas and logic accurately. I believe this is due to their poor English and lack of adequate vocabulary to describe the abstract science concept.
>
> (T12)

Another late–partial EMI biology teacher reported that teachers would help students to interpret questions and then help them to memorise the answers and write the answers once by themselves. She added, 'The grammar is not important as long as they apply the right concepts and flow of ideas.'

Ms. Leung, a late–partial EMI physics teacher, also added that one major concern is that students who attempted to express ideas in their own words often failed to provide the correct explanation. She explained that 'this is due to students' poor answering skills. I can just ask them to answer in detail'. She approached this problem by encouraging students to highlight key points and cut long sentences into short phrases in order to simplify the abstract and lengthy science content. She said:

> For English students, I will need to take more care of them, since their English ability is relatively low. I have to highlight the key points in the textbook in order to draw their attention to the key words and also to simplify the content needed for memorization … Highlight the conceptual words in questions, hope they can recognise these concepts and be able to understand the questions in the future, in order to have a greater chance to get the marks.
>
> (T8)

On rote memorisation, a late–partial EMI chemistry teacher stressed that weaker students would not spend time unpacking the technical, dense, abstract science expressions. Instead, they would memorise lengthy sentences to describe a science phenomenon in examinations. As Mr. Yu, a late–partial chemistry teacher, explained:

> Students find it difficult to use specific terms to explain a concept. Instead, they just resort to lengthy sentences. Some students don't understand specific concepts and even have wrong concepts, which may be due to their lack of reading textbooks or practising practically. Others simply can't express systematically.
>
> (T7)

Similarly, Ms. Leung, a late–partial biology teacher, noted:

> Students directly copy everything from the textbook when answering questions. This is because students get used to the learning style at junior forms … I would remind students and encourage them to critically analyse the science phenomena.
>
> (T8)

Overall, late–partial EMI teachers thought that this attitude towards language use was likely to create significant problems when students attempted to explain science concepts.

A few biology teachers expressed concerns about their students handling the new biology curriculum. For instance, Mr. Wong, an early–full biology teacher, said of his students:

> they cannot interpret exam questions accurately; cannot handle important concepts, although these concepts have been questions in public exams for many years; cannot handle abstract questions related to the nature of science. Students' English is not too bad. It's a lack of student motivation to memorise vocabulary and phrases.
>
> (T4)

Other biology teachers from both groups expressed concerns about students' communication skills in presenting science ideas in writing and in speaking. Students could not present coherent science explanations in English with adequate vocabulary and expressions. Mr. Ng, a late–partial biology teacher, explained:

> I think it is the part they performed the worst. They can't express what they want. Firstly, the technical terms are too difficult, so they don't know how to use them. Secondly, students know the terms in Chinese, but don't know how to present them in English. Thirdly, they don't get used to speaking

English. In classroom interaction, students are not so willing to use English to raise questions. I think it is because they don't have many chances and the environment to practise English.

(T15)

Another late–partial biology teacher, Mr. Law, said:

The biggest problem is students are bad at presenting these ideas in exams and tests. This problem is related to their English. They understand the ideas, but since they use inaccurate vocabulary or write incomplete answers, they can't get the marks. Sometimes when it comes to long questions [questions that require comparing and contrasting and systematic explanation], I can get their points, but their answers are just not coherent. They have so much information, but they don't know how to organise it into a logical argument or explanation.

(T9)

In sum, teachers from both schools highlighted language challenges regarding: students' lack of analytical skills necessary to understand the examination questions, identify the important concepts or interpret abstract questions; students' lack of motivation to read extensively about the learnt science concepts in textbooks or other sources like the internet; students' lack of communication skills in presenting science ideas in a logical, coherent way with the appropriate use of scientific vocabulary and expressions; students' use of rote memorisation; and students' failure to express science ideas with the correct English vocabulary.

6.2.3.3   *Coping strategies*

Teachers from both groups indicated that they used a range of pre-class activities to accommodate students. Mr Lui (T2), an early–full EMI physics teacher, said that the use of pre-lab-experiment videos could raise students' awareness of the steps necessary in the course of conducting a lab-experiment. He would ask his students to comment on what was done in the video and consider what precautions they would need to consider when carrying out the experiment. He found that this motivated students to become more involved, challenged their existing knowledge of science concepts and improved their analytical skills in understanding an unfamiliar situation. Mr Yu (T9), a late–partial chemistry teacher, would encourage students to read aloud the questions in class. This strategy helped students explain how to address different questions by identifying the specific question types, such as 'describe' or 'account for'. Mr. Yu, a late–partial chemistry teacher, explained:

When coming across some difficult terms, I will ask them to read them aloud. What's more, they are required to practise past papers and I will teach them how to answer specific question types.

(T9)

Late–partial EMI science teachers would encourage students to practise writing explanations for long questions, thereby helping students memorise the correct scientific expressions. Mr. Law (biology, late–partial) hoped that this strategy would help students develop their knowledge of scientific expression in writing:

> I will provide more training to my students by asking the students to prac-
> tise how to answer long questions for homework. Students usually perform
> badly, because the topic is new to them and they have not done any revision
> yet … I encourage my students to spend time on memorising the adequate
> scientific expressions; that really helps them to construct an articulated sci-
> ence argument in a logical manner and to connect different technical terms.
>
> (T9)

Three early–full teachers (2 physics and 1 chemistry) suggested using various technologies, such as flipped-classroom or text-messages, to improve the quality of teaching and hence address students' language challenges. For example, Mr. Lui, an early–full physics teacher, recommended that students learn all the key concepts at home by watching a teacher-prepared video and come back with questions to be discussed in class. He said that this flipped-classroom model can create more opportunities for meaningful interaction, and more time for students to ask questions. As students all watch the same video at home, teachers can ensure that the teaching process is consistent across different classes. This relaxes the teaching process, reducing the time necessary for preparing classroom teaching materials. He explained:

> My teaching is more relaxed, as I don't have to shout and students aren't
> able to disturb my teaching. Also, writing on the blackboard, showing the
> PowerPoint and waiting for students to copy notes can waste quite a lot of
> time during lessons. By using this method, if students can't follow, they can
> just pause the video, and it will waste less time during the lesson.
>
> (T2)

Mr. Luk also used the flipped-classroom model, recognising that it allowed students to:

> do more exercises in class, and generate more discussion around the topic.
> I see it is an active way of learning, as they could discover what they don't
> understand or lack. At the same time, they can ask me (teacher) questions and
> discuss with other classmates immediately to clarify their misconceptions.
>
> (T2)

Teachers also reported using other technologies. Mr. Lai, an early–full physics teacher, found that an e-learning platform similarly increased classroom capabilities by allowing students to learn first at home. He noted:

An e-learning platform allows students to learn at home and review the teaching materials several times. Students have more questions to ask now. They ask questions almost every day. Students would have more chance to practise their science knowledge via the assignments provided by the e-learning platform.

(T3)

He further explained:

Online teaching video clips are provided and I (teacher) will know whether students have watched or not because there is a record. In addition, if students have asked me questions but the answers have already been mentioned in the video, I will know that s/he hasn't seen the video, but of course, if they still have questions, I will go through the video with them.

(T3)

Another late–partial EMI physics teacher, Mr. Lai, used an e-learning platform during the class to help students to learn physics concepts by interacting with the online learning materials. He said that the e-learning platform could allow students to continue their learning at home and accommodate individual learning differences. The platform allowed him to check students' progress in learning essential science concepts. He also added that students could overcome their language challenges in English by watching the Chinese video first to understand the main concepts and then watch the English version to develop knowledge of English expressions. He saw this as a means to help students effectively learn both content and language.

Mr. Lai (early–full, physics) also provided online learning materials in both Chinese and English versions, recognising that:

Students can watch the online teaching videos that are presented in Cantonese and in English. Students can choose to watch it in Cantonese, and later in English to ensure they understand science concepts in Cantonese and later the technical terms in English. Students would be reminded to highlight key points in the textbooks and draw simple diagrams to clarify the concepts. I would provide more exercise to help students to practise.

(T3)

Several teachers used a mobile text messenger service to facilitate interactive discussions. Mr. Lai said that through this platform:

students can raise questions any time they like, and they can share YouTube video clips that are related to Physics on WhatsApp group and discuss them together.

(T3)

Mr. Yu (late–partial chemistry) used WhatsApp, but found it only worked for students that were already highly motivated:

> I have set up a WhatsApp group for my students to raise their problems. Students with good results are highly motivated and interactive as they keep asking and answering questions in English but there is no response for students of poor academic results.
>
> (T9)

Physics teacher Mr. Lai found that his interaction with students increased significantly after he started to communicate with them using a mobile online text messenger. This tool helped build rapport and relationships with students, allowing students to ask questions at home and share questions and answers with other students. Teachers could also distribute relevant science materials online and create a discussion among students to cultivate their interest in learning extended topics in science.

6.2.3.4   *Specific coping strategies: examples from biology class*

A number of biology teachers shared coping strategies used to address students' language issues in studying biology. One biology teacher, Mr Wong, described the nature of study in secondary biology as learning a great deal of vocabulary and their semantic relations in different biological contexts. He explained:

> I think teaching is a way of living. I don't separate different types of vocabulary to teach. Our goal is not making students understand that three kinds of vocabulary. Our goal is teaching students to read and write, and then get marks. You will mention technical terms, comparison and keywords naturally while you are teaching a topic.
>
> (T4)

A biology teacher from the early–full EMI schools suggested six strategies to promote effective learning. They include:

- Using examples from daily life to demonstrate what students have learnt.
- Asking questions to prompt students' answers, which leads to constructing science explanations together with the students by making summaries.
- Asking more higher-order questions to check students' understanding of the cause-and-effect relationship between different factors and variables of a science phenomenon. This strategy can help students to understand the principle of the experiment.
- Checking students' misconceptions (both in lab and lectures) that are common mistakes in exams (e.g., the difference between sodium hydrogen carbonate and sodium carbonate).
- Connecting students' biological concepts with other science subjects (e.g., reminding students of the properties of water they have learnt about in chemistry when teaching the concept of diffusion in biology).

- Drawing simplified diagrams on the blackboard to demonstrate the steps for describing experimental procedures and explaining experiment results. (Through graphic presentations, students can more easily understand the entire process of scientific investigation, and at the end will be able to use words to present the scientific explanation.)

In sum, teachers from the early–full EMI schools suggested a number of strategies they use to help students solve language challenges. They include:

1 highlighting key words and dictations
2 going through past papers (sentence structure in the marking scheme)
3 understanding questioning words
4 going through similar lexical items (e.g., anti-words like antibiotics and anti-body)
5 using multiple-choice questions to consolidate science concepts
6 watching the interactive/lab-experiment videos to arouse students' interest in related science topics
7 reading aloud in class to familiarise the vocabulary and the text
8 practising scientific expressions.

6.2.3.5   *When to use L1*

Teachers in the early–full EMI schools were asked in what situation they would opt for L1. If the school policy allows a mixed-mode instruction, these teachers would be likely to use L1 in the classroom and encourage students to use Chinese when they found it difficult to express themselves in English. They found that this strategy often worked for students with limited English proficiency. Teachers reported asking their students to translate what the teacher asked into Cantonese and then allowing them to respond in Cantonese. Once students developed confidence in the content, the teacher would then guide them to respond in English through closed questions, thereby reducing the demand on students' output. They would gradually encourage them to produce phrases and short utterances in English with open-ended questions. These teachers felt that this mixed-mode instruction helped students learn English more effectively. Mr. Wong, Mr. Lai, and Mr. Wong, respectively, reported:

> If they really can't express their ideas in English, I encourage them to use Chinese, because I just want to check whether they understood. Sometimes if they don't understand what I am asking, I will repeat the question in Chinese as well.
>
> (T1)

> I use Cantonese when I communicate with students with lower ability individually, rephrasing what the student said and trying to correct his wording in English. I provide hints, key words to students, so that they can continue their answer.
>
> (T2)

Students tried to use Cantonese to answer my questions. Firstly, I remind them to try to use English. Secondly, I give them some guiding questions, turning the question from open-ended to close-ended. Thirdly, I give students choices and ask them to pick the right answer.

(T5)

Teachers in both early–full and late–partial schools would also consider using L1 in their teaching when explaining abstract ideas and important steps in experiments, as sometimes the experiment could be dangerous. They would also use L1 in giving examples or when communicating individually with students with lower ability. Ms Chow, an early–full biology teacher, thought that 'an interactive atmosphere is of vital importance' as it could help to promote the acquisition of scientific knowledge and language. She noted that:

The learning style of each student varies. Some prefer lecturing and doing exercises while others prefer interaction. What's more, different students have different ways to express their thoughts. Some do better in verbal communication and others in writing. Apart from that, students' attention span is limited, so they will get bored if I keep speaking.

(T8)

She also added that she would encourage students to speak in Cantonese first and check their understanding. For those who have problems in articulating their ideas in English, she would encourage them to translate it from Cantonese to English:

If they have questions for me, I won't reply directly. Instead, I will allow them to generate the answer first.

(T7)

In this way, Ms Chow believes that students can build up their confidence and clearly understand the reasoning behind the content, which requires higher-order thinking, which is difficult for students to do in English. Quick mediation through Cantonese can help students to develop long-term memory and reasoning. Other science teachers from the late–partial schools would explain the English expressions and draw students' attention to different meanings of certain writing styles. For example, one physics teacher paid attention to the specific meaning of certain words and explained to students, e.g., that 'what time' in physics is actually asking the duration, and that this meaning is therefore different from everyday English.

The teacher interviewees also suggested a number of ways to deal with students who have difficulty in communicating in English. Three science teachers in the early–full EMI schools reported that they collaborated with English language teachers at their school to improve students' abilities in reading and writing science through English. They communicated with English teachers about their students' language weaknesses (e.g., problems with question words, writing expressions), hoping the English teachers would revisit these language problems in English lessons. Two science teachers in the early–full EMI schools became involved in the Language Across Curriculum class in junior forms to develop students' scientific language competence from junior to senior levels.

6.2.4   *Summary of the teacher interviews*

The above analysis has explored the views of science teachers in early–full and late–partial EMI schools on their teaching experiences in EMI science classrooms. In the semi-structured interviews, the participating teachers were asked about their perceptions of EMI teaching in science classrooms.

Eight recurring themes were identified from the teachers' interview transcripts: (1) teacher views on teaching experience in EMI science classrooms; (2) language challenges in teaching science through English; (3) coping strategies; (4) dealing with students with difficulty in communicating in English; (5) when to use L1; (6) questioning techniques in EMI classrooms; (7) vocabulary types in EMI science classrooms; and (8) other teaching challenges related to the new science curriculum. As the discussion indicated, the teachers in both the early–full and late–partial EMI schools shared similar views about numerous aspects of their EMI teaching experiences, which I summarise below.

6.3   Views shared by teachers in early–full and late–partial schools on English as the MOI

1   Many science teachers reported that they are not good at English themselves (none of the early–full and late–partial teachers who participated in the interviews agreed that they speak good English as indicated in their lower score in their self-reported English speaking) and thus use Chinese in addition to English in their classrooms. They generally welcome a mixed-mode instruction with Cantonese as the matrix language embedded with English scientific terms. The use of L1 allows teachers to easily explain logical relationships, provide comprehensive explanations to students with limited English, and consolidate students' knowledge by making connections with examples from daily life.

2   Students' analytical skills, reading skills, and communication skills are the major concerns highlighted by science teachers. Students have problems in presenting science ideas in writing and speaking. Students may not understand examination questions, identify important concepts, or interpret abstract questions. They cannot present coherent science explanations in English with adequate vocabulary and expressions.

3   Teachers suggested that the use of techniques, such as a flipped-classroom, e-learning platforms, online materials with Chinese and English versions, and text messengers, improve the quality of teaching and hence address students' language challenges.

4   When asked about when teachers should use L1, the majority of late–partial teachers said they used L1 for students with limited English abilities and when

referring to everyday examples that might be common in Chinese. Some of these teachers preferred to use L1 first and then recap in L2, particularly in the following situations: when explaining abstract ideas; when explaining important steps of dangerous experiments; when explaining daily life examples only known in Chinese; or when communicating with students with lower ability individually.

5   Teachers generally used a range of question types, such as recalling, higher-order, and reasoning, to promote students' science learning. Preference for types of question was different between groups. Teachers in early–full EMI schools frequently used recalling questions to remind students of what they have learnt previously. Teachers in late–partial EMI schools used more reasoning questions to develop students' skills in reasoning and critical thinking. One possible reason for the different preferences is that late–partial EMI teachers mediated through Cantonese, avoiding the use of English and/or technical terms, and thus focused more on the reasoning behind science concepts. Teachers in both groups used reasoning questions in students' discussions where students could provide their answers in short English phrases.

6   One biology teacher in early–full EMI schools suggested a sequence of questioning: start by using recall questions to remind students of the factual information and involve them in the class interactions. If they grasp the concepts, follow with higher-order questions which aim to increase student–student interaction and discussion on key concepts. Later, supplement with reasoning questions. The teachers felt that through this technique, the students' cognitive thinking can be stimulated.

7   There are three types of vocabulary that students are likely to encounter in learning science: basic English vocabulary, academic vocabulary, and scientific or technical vocabulary. Students had more problems with basic English vocabulary than technical vocabulary because they misinterpreted their meanings in science and confused them with other meanings in English lessons. Students also mixed up the meanings of some technical vocabulary in different science contexts

8   Teachers believe that students' poor vocabulary affected their writing ability, especially when students are asked to write an essay to explain a science phenomenon and particularly in biology. Some teachers suggested techniques such as drawing a concept map and/or producing short paragraphs with logical and coherent ideas and then connecting a number of short paragraphs into a structured essay.

9   Teachers also believe that poor vocabulary affected students' communication skills.

10 When asked about the expectation of students' science achievements in EMI classrooms, the majority of science teachers hoped their students would master the key science concepts, develop strong observational skills, improve their analytical and critical thinking skills, and ultimately show a strong interest in science and its application in our society.

6.4   Points of difference between early–full and late–partial teachers on English as the MOI

1 Teachers in the late–partial EMI schools strongly believe the medium of instruction is merely a tool for transmitting knowledge, rather than a means for preparing students to learn a second language, while teachers from the early–full EMI schools stressed that English is the only preferred language because most of the scientific terms are from English, e.g., teaching the periodic table, labelling organic compounds.
2 Teachers from the early–full EMI schools pointed out that students faced a number of language challenges when learning biology through English. They expressed their concerns over their students' analytical skills, reading skills, and communication skills. One major concern is that students who attempted to express themselves in their own words often failed to provide a correct explanation.
3 In the late–partial EMI schools, teachers said that rote memorisation was problematic for weaker students who did not spend time unpacking the technical, dense, abstract science expressions. These students only memorised lengthy sentences in order to describe a science phenomenon in examinations.
4 Teachers in the early–full EMI schools suggested a number of methods to help students solve their language challenges by making the English more accessible to students with low English ability. These methods include highlighting key words and dictations, understanding questioning words, and going through similar lexical items (e.g., anti-words like antibiotics and anti-body).
5 When dealing with students who have difficulty communicating in English, the two groups had different approaches. A few teachers in early–full EMI schools reported that they collaborated with English language teachers at their school to improve students' abilities in reading and writing. The majority of teachers, particularly in the late–partial EMI schools, used Cantonese in teaching and encouraged students to explain in L1.

Teachers from both groups also reported a number of concerns and teaching challenges related to the new science curriculum. They shared their views on: (1) teaching students with diversified science backgrounds; (2) overloaded teaching, too many topics; (4) rearrangement of teaching topics according to students' level; (5) overloaded curriculum and limited study-time; (6) declined status of science subjects in the new curriculum; and (8) gender making a difference in selecting the MOI for public examinations. The teachers in both the early–full and late–partial EMI schools had both similar and differing views.

6.5   Summary

This chapter has reported on the teachers' perceptions of their EMI experiences in both early–full and late–partial EMI schools. The correlation analyses showed that a number of teachers' demographic characteristics correlated with the teachers' perceptions of their EMI experiences. The descriptive analyses revealed differences of teachers' perceptions between early–full and late–partial EMI schools. Similar patterns were observed in the results of early–full and late–partial EMI students' perceptions of their EMI experiences (discussed in Chapter 8). The qualitative interview analyses demonstrated that the early–full and late–partial teachers shared a similar view about many aspects of their EMI teaching experiences. However, when dealing with students with difficultly in communicating in English, the two groups of teachers used different approaches. Under the Education Bureau's initiation with more emphasis on language across curriculum (Education Bureau, 2017), a few interviewed early–full EMI teachers reported having collaborations with English language teachers at their schools to improve students' abilities in reading and writing science through English. The late–partial EMI teachers would use optimal Cantonese in teaching and encourage students to explain in L1. Further, both groups generally used a range of question types to promote students' science learning but the preference is different between groups. Teachers in early–full EMI schools used recalling questions to remind students of what they had learnt previously, while late–partial teachers preferred to use reasoning questions (i.e., mediating through Cantonese) to develop students' skills in reasoning and critical thinking. In the next chapter, I will present the analysis of surveys and interviews with students about their EMI experiences. A total of 545 students in early–full and late–partial EMI schools will be compared in terms of their choice of instructional language for science teaching activities, their views on the most demanding and important English language skills, their teaching beliefs about EMI classrooms, their language awareness, and the language challenges and coping strategies.

Reference

Education Bureau (2017). *Medium of instruction for secondary schools sharing session - a whole-school approach to language across the curriculum* (24 November 2012). https://www.edb.gov.hk/en/edu-system/primary-secondary/applicable-to-secondary/moi/support-and-resources/experience-sharing-sessions/read-fair-2012_20121124.html

7 Science students' perceptions of their EMI experiences

7.1 Students' perceptions of their learning experiences in EMI science classrooms (questionnaire data)

This section presents a detailed analysis of data and findings about the students' perceptions of their English medium instruction (EMI) learning experiences in science classrooms based on their questionnaire responses.

7.1.1 Demographic characteristics of the student respondents

A total of 545 students in early–full and late–partial EMI schools in Grades 10 and 11 returned their questionnaires, with a total response rate of 90%. The questionnaire collected information pertaining to the students' demographic characteristics and consisted of six aspects which explored the students' experiences in EMI science classrooms: (1) their preference of instructional language in their current science classrooms, (2) the most important and demanding English language skills needed to study science in English, (3) their reflections on science learning in the EMI context, (4) their self-concept in learning science, (5) the language challenges they encountered when learning science in English, and (6) the usefulness and frequency of the coping strategies that they used. Table 7.1 shows the demographic characteristics of the participating students.

7.1.2 Checking the internal consistency of survey items

The survey data measuring students' perceptions of their EMI experience and their internal consistency are displayed in Table 7.2. Cronbach's alpha coefficient is smaller than 0.7, suggesting that the items have relatively low internal consistency. However, given that the three items related to students' reflections on the impact of MOI on their science learning are central to my research questions, and that students from different MOI schools could have diversified views on their reflections, I kept this construct for the rest of the analysis.

DOI: 10.4324/9781003001454-8

Table 7.1 Characteristics of the student respondents

Characteristics		Students who returned questionnaires (n = 545); response rate = 90%	
Group, *n*, (%)		Early–full EMI *n* = 284 (52%)	Late–partial EMI *n* = 261 (48%)
Grade	10	106	144
	11	178	117
Sex	Male	101 (19%)	136 (25%)
	Female	182 (33%)	125 (23%)
Age	15	0	40 (7%)
	16	56 (10%)	102 (19%)
	17	150 (28%)	88 (16%)
	18	59 (11%)	24 (4%)
	19 or above	19 (3%)	7 (1%)
First language	Cantonese	282 (52%)	258 (47%)
	Mandarin	2 (0.4%)	3 (1%)
Place of primary education	Hong Kong	282 (52%)	258 (47%)
	Mainland China	2 (0.4%)	3 (1%)
Chinese language achievements from last year's school exam		3.22 (1.06)	3.11 (0.91)
English language achievements from last year's school exam	Listening	2.73 (1.02)	2.61 (0.90)
	Speaking	2.67 (1.05)	2.75 (0.844)
	Writing	2.8 (0.98)	2.77 (0.87)
	Reading	2.75 (0.98)	2.54 (0.84)
Science achievement from last year's school exam		2.9 (1.02)	2.49 (0.79)

7.1.3   Checking normality

The means of the responses to the items in the student questionnaire were calculated and all the variables were explored with reference to normality of distribution using the Kolmogorov–Smirnov test. Any value of the Kolmogorov–Smirnov (K–S) Z greater than 0.05 indicates normality. If the significant value is lower than .05, it suggests violation of the assumption of normality. Accordingly, the final decision was to use a parametric test (independent *t* test) for scales 2, 4, 5, and 6, and to use a non-parametric test (Mann–Whitney) for scales 1, 3, 7, 8, and 9.

7.1.4   Students' perceptions of EMI experiences and demographic characteristics

A Spearman's rank-order correlation was run to determine the relationship among the 16 demographic characteristics of the students and their responses to the

Table 7.2 Cronbach's alpha reliability for the five scales of questionnaire variables

Code	Description	Cronbach's alpha (α)
1. **Language choice in science activities**	Choice of instructional language in different science teaching activities (13 items)	0.922
2. **English demanding**	The most demanding English language skills for students to study science in English medium (17 items)	0.928
3. **English important**	The most important English language skills for students to study science in English medium (17 items)	0.955
4. **EMI reflect**	Student's reflections on the impact of MOI on their science learning (3 items)	0.519^
5. **Self-concept**	Student's self-concept in learning science through English (7 items)	0.781
6. **Belief**	Student's beliefs about instructional activities in learning science through English (9 items)	0.722
7. **Challenges**	Students' views towards the language challenges for students when learning science through English (27 items)	0.949
8. **Frequency strategy**	Students' views towards the frequency of the coping strategies (9 items)	0.759
9. **Usefulness strategy**	Students' views towards the usefulness of coping strategies (9 items)	0.821

questionnaire items relating to their perceptions of EMI experiences. The results with statistical significance are as follows:

- Seven students' demographic characteristics were strongly correlated with the students' choices of instructional language in science teaching activities. These were school's current MOI ($r = 0.270$, $p < 0.0001$), previous MOIs in primary ($r = 0.385$, $p < 0.0001$) and junior secondary ($r = 0.247$, $p < 0.0001$) schools, students' English language overall results in their previous school year ($r = 0.162$, $p < 0.0001$), and English productive skills of speaking ($r = 0.182$, $p < 0.0001$) and writing ($r = 0.149$, $p < 0.0001$). This perhaps suggests that students who tend to use Cantonese to assist their science learning activities are likely from late–partial EMI schools, with many years of Cantonese instruction in their primary and junior secondary years, and perhaps have relatively higher scores for their English productive skills (speaking and writing), because they have used Cantonese to help them develop their understanding of certain concepts. Other variables such as sex and science subjects were negatively correlated with the students' choices of instructional language. Students' receptive English skills (reading and listening) were found to have a low and positive correlation. This may suggest that male students and those who study physics tend to use more Cantonese in their science learning activities. Likewise, students with higher scores for reading and listening tend to use more Cantonese.

- Five students' demographic characteristics were strongly correlated with the students' views on their most demanding English language skills. Students' self-reported overall English language results in their previous school year ($r = 0.249$, $p < 0.0001$) and individual English skills—speaking ($r = 0.192$, $p < 0.0001$), reading ($r = 0.21$, $p < 0.0001$), writing ($r = 0.206$, $p < 0.0001$) and listening ($r = 0.240$, $p < 0.0001$)—were found to have a strong and positive correlation with the students' views on their most demanding English language skills. This perhaps suggests that students who reported lower English language results in previous school years rated the most demanding English language skills higher. Other variables such as science subject were negatively correlated, whereas previous MOI in primary levels was positively correlated. This may suggest that students who study physics and use more Cantonese at primary levels rated the demanding English language skills higher.
- The most important English language skills needed for students to study science through the medium of English were found to have a strong and negative correlation with the students' results in writing English ($r = -0.152$, $p < 0.0001$). This perhaps suggests that students who reported a low score for writing rated the importance of the English language skills higher. Other variables such as current MOI, students' overall English scores and listening scores in the previous school year were found to have a negative correlation with students' views on the importance of English language skills. This suggests that students from early–full EMI schools, and those with higher English scores tended to rate the importance of the eight English language skills higher. Sex and age were found to have a low and positive correlation. This suggests that female and older students rated the importance of the eight English language skills higher.
- Four variables were found to have low and negative correlations with the student's reflections on EMI. They were students' previous MOI at the junior level ($r = -0.099$, $p = 0.02$), overall English scores in the previous school year ($r = -0.106$, $p = 0.013$), English speaking score ($r = -0.112$, $p = 0.009$) and writing score ($r = -0.115$, $p = 0.007$). This perhaps suggests that students who rated their EMI reflection higher experienced more English instruction in their previous MOI at junior secondary but also had lower results in their overall English score and lower results in speaking and writing.
- Four variables were found to have a negative correlation with students' self-concept in science. These were students' sex ($r = -0.136$, $p = 0.001$), students' science score ($r = -0.291$, $p < 0.0001$), and students' English reading score ($r = -0.106$, $p = 0.014$) and writing score ($r = -0.089$, $p = 0.037$). This perhaps suggests that male students with lower scores for science, as well as English reading and writing had a higher level of self-concept in science than female students with higher scores for science and English reading and writing.
- Nine variables were found to have a negative correlation with students' beliefs in science classrooms. They were students' age ($r = -0.109$, $p = 0.011$), previous MOI in primary ($r = -0.115$, $p = 0.007$) and in junior secondary ($r = -0.148$, $p = 0.001$), overall English score ($r = -0.103$, $p = 0.016$), reading ($r = -0.111$, $p = 0.01$), writing ($r = -0.088$, $p = 0.04$), listening ($r = -0.115$, $p = 0.007$), Chinese language ($r = -0.112$, $p = 0.009$), and science ($r = -0.166$, $p = 0.001$).

This seems to suggest that older students, with more Cantonese instruction in their primary and junior secondary levels, and higher scores for English, Chinese, and science, had a lower level of student belief regarding whether they had the ability to do well in science classrooms.

- Nine variables were found to have a correlation with the students' views on language challenges. A positive correlation was found with current MOI ($r = 0.157, p < 0.0001$), previous MOI in primary ($r = 0.146, p = 0.001$) English score ($r = 0.209, p < 0.0001$), speaking ($r = 0.178, p < 0.0001$), writing ($r = 0.162, p < 0.0001$), and listening ($r = 0.225, p < 0.0001$). This perhaps suggests that students from late–partial EMI schools with higher scores for English speaking, writing, and listening rated the language challenges higher. A weak and negative correlation was found with students' current grade ($r = -0.105, p = 0.014$) and students' age ($r = -1.35, p = 0.002$). This may suggest that younger students in lower grades (i.e., Grade 10) rated the same set of language challenges higher.

- Four variables were found to have a positive correlation with the students' views on the frequency of using coping strategies. They were students' age ($r = 0.12, p = 0.005$), previous MOI in primary ($r = 0.124, p = 0.004$), English reading score ($r = 0.105, p = 0.015$), and listening score ($r = 0.105, p = 0.015$). This appears to suggest that older students with more Cantonese as the MOI at primary levels, as well as those with higher scores for reading and listening, used coping strategies more frequently.

- Nine variables were found to have a correlation with the students' views on the usefulness of coping strategies. A strong and positive correlation was found with students' age ($r = 0.241, p < 0.0001$), previous MOI at primary levels ($r = 0.190, p < 0.0001$), overall English score ($r = 0.124, p = 0.004$), reading score ($r = 0.149, p < 0.0001$), listening score ($r = 0.147, p = 0.001$) and science score ($r = 0.138, p = 0.008$). This seems to suggest that older students with more Cantonese at the primary level, and with higher scores for English and science, rated the usefulness of coping strategies higher. A negative and strong correlation was found with the current MOI at school ($r = -0.256, p < 0.0001$) and science subject ($r = -0.179, p < 0.0001$). This may suggest that students from the early–full EMI schools who choose biology as their science subject rated the usefulness of coping strategies higher.

7.1.5   *Correlations among the nine questionnaire scales*

A Spearman's rank-order correlation was run to determine the relationship among the nine scales of the questionnaire. The results with statistical significance are as follows.

7.1.5.1   *Choice of instructional language in different science teaching activities*

The language choice in science teaching activities was found to have strong and positive correlations with students' rating of the demanding English skills in learning science ($r = 0.152$, $p < 0.0001$), which refers to whether students believed learning science through English was exacting for them, along with the language challenges of learning science in English ($r = 0.212$, $p < 0.0001$). Similarly, it was found to have a strong and negative correlation with students' rating of the importance of English skills in learning science ($r = -0.126$, $p = 0.003$), students' belief in learning science ($r = -0.182$, $p < 0.0001$), and the usefulness of the coping strategies used by students ($r = -0.107$, $p = 0.012$). This perhaps suggests that students with more Cantonese or code-switching in the science teaching activities rated their most demanding English skills as well as the language challenges of learning science in English higher. Students engaged in more English-based science activities rated the following factors higher than their peers: importance of English skills, belief in their ability in learning science and the usefulness of coping strategies.

7.1.5.2 *The most demanding English language skills for students to study science in English medium*

The demanding English skills category, which measured whether students thought learning science through English was exacting for them, was found to have strong and positive correlations with students' views on language challenges ($r = 0.560$, $p < 0.0001$). There were moderate and positive correlations with the students' views on the frequency of using coping strategies ($r = 0.116$, $p = 0.007$) and the usefulness of the coping strategies ($r = 0.129$, $p = 0.003$). Further, there were weak and negative correlations with the students' self-concept in science ($r = -0.093$, $p = 0.031$) and with the students' beliefs in science classrooms ($r = -0.092$, $p = 0.032$). This suggests that students who rated their demanding English skills higher also rated the language challenges and the frequency of using coping strategies higher. Those who rated their demanding English skills lower tended to have a higher level of self-concept in science and a stronger belief in science classrooms.

7.1.5.3 *The most important English language skills for students to study science in English medium*

The importance of English language skills was found to have a strong and positive correlation with the students' beliefs in science classrooms ($r = 0.191$, $p < 0.0001$) and with the usefulness of the strategies used by students ($r = 0.206$, $p < 0.0001$). There was also a strong negative correlation with the language choice in science activities ($r = -0.126$, $p = 0.003$). This perhaps suggests that students who rated their important English skills higher had a stronger belief in science classrooms and found the coping strategies more useful. They also tended to use more English in science teaching activities.

7.1.5.4 *Students' reflections on EMI*

The students' reflections on the EMI category was found to have a strong positive correlation with students' self-concept in science ($r = 0.261$, $p < 0.0001$) and with students' beliefs in science classrooms ($r = 0.161$, $p < 0.0001$). This category was also found to have a strong and negative correlation with language challenges of learning science ($r = -0.138$, $p = 0.001$). This perhaps suggests that students who rated their reflection on EMI higher had a better self-concept in science and stronger belief in science classrooms. They also tended to use more English in the science teaching activities.

7.1.5.5   Students' self-concept of science

The students' self-concept of science category was found to be strongly positively correlated with students' beliefs in science classrooms ($r = 0.305$, $p < 0.0001$) and strongly negatively correlated with language challenges of learning science ($r = -0.197$, $p < 0.0001$). This perhaps suggests that students with stronger self-concept in science had a stronger belief in science classrooms and that they tended to use more English in science learning activities.

7.1.5.6   Students' views on language challenge

The students' views on language challenges category was found to have a positive correlation with the frequency of using coping strategies ($r = 0.110$, $p = 0.01$) and usefulness of coping strategies ($r = 0.096$, $p = 0.025$). This suggests that students with more language challenges used more coping strategies and found them more useful.

7.1.5.7   Usefulness of coping strategies

The students' views on the usefulness of coping strategies category was found to have a positive correlation with the language challenges category ($r = 0.601$, $p < 0.0001$). This suggests that students with more language challenges found coping strategies more useful.

7.1.6   Summary of the correlations

The major findings of the correlations among different aspects of the students' perceptions of their EMI experience are as follows:

1 Students who used more Cantonese in the science teaching activities likely struggled more with demanding English skills and language challenges.
2 Students who used more English in their science learning activities likely saw the importance of English skills, believed in learning science, and found coping strategies useful.
3 Students who found English skills more demanding were likely to report more language challenges and a higher frequency of using coping strategies.
4 Students who found English skills less demanding were more likely to have a higher level of self-concept in science and a stronger belief in science classrooms.

5 Students who rated the importance of English skills higher were more likely to have a stronger belief in science classrooms and find coping strategies useful. These students tended to use more English in science activities.
6 Students who rated their reflections on EMI higher had a better self-concept in science and a stronger belief in science classrooms. These students tended to use more English in science activities.
7 Students with a stronger self-concept in science also had a stronger belief in science classrooms and tended to use more English in science activities.
8 Students with more language challenges used more coping strategies and found them useful.
9 Students who used coping strategies more frequently also found them useful.

The results of the correlation analysis suggest that there are three possible cases of students in both early–full and late–partial EMI schools.

1 Combining the two groups, students who prefer to use more Cantonese in interactions find that English language skills are more demanding and experience more language challenges in science classrooms. Thus, these students frequently use coping strategies. These students are possibly poorly performing students in the early–full EMI schools or late–partial EMI schools, but the questionnaire analysis showed that the late–partial students used coping strategies less frequently than the early–full students.
2 Students who perceive English language skills as important also believe they can do science well. These students also tend to prefer to use more English in their interactions and find coping strategies more useful. This is possibly the case for well-performing students in both early–full and late–partial EMI schools.
3 Students who have more language challenges use coping strategies more frequently and find them useful. This is possibly the case for students who are struggling with English but motivated to overcome their language challenges in both early–full and late–partial EMI schools.

7.2   Early–full EMI vs. late–partial EMI

Whether there were any difference in students' perceptions of their EMI experiences in science classrooms between students from early–full EMI schools and students from late–partial EMI schools? To answer this question, I first conducted principal component analyses to reduce the number of survey items from the nine questionnaire scales. I then compared the mean differences of the total students' responses between the early–full EMI and late–partial EMI students in each of the questionnaire scales. The first factor comprised the following 17 language challenges, listed in order of heaviest to lightest and worded the same as they appeared in the questionnaire.
    Factor 1:

1 The teaching methods of an individual teacher
2 Understanding the meaning of technical terminology in English

3 Understanding the course content in English
4 Explaining a science concept in spoken English after reading the textbook
5 Comprehending the acquired science knowledge
6 Discussing with other classmates about science in English
7 Explaining an abstract science concept in English
8 Understanding my science teacher's written English during lessons
9 Explaining a science concept in written English after reading the textbook
10 Explaining a difficult science concept in English
11 Answering a question from my science teacher in English
12 Understanding my science teacher's spoken English during lessons
13 Application of the acquired science knowledge
14 Studying to meet teachers' expectation in a science topic
15 Distinguishing the basic English vocabulary and technical terminology
16 Reading my English science textbook
17 Finding the appropriate Chinese glossary for reference.

All 17 of these language challenges suggest that students in EMI classrooms experienced difficulties in understanding, comprehending, and communicating science content through English. The most heavily weighted challenge is about the teaching methods of an individual teacher, which is less direct about the students' own language challenges. The next two most heavily weighted challenges are related to understanding and comprehending scientific terms and content materials in English. All 17 language challenges are closely related to the concept of cognitive academic language proficiency skills (CALP). CALP refers to the language skills that are essential for students to succeed in schools and these skills are expected to be demonstrated in content-subject areas through formal academic learning (Cummins, 2000). As a result, this component was labelled as 'challenges (use of scientific English)'.

The second factor was made up of seven language challenges
Factor 2:

1 Expressing the right answers in written English
2 Solving problems in written English
3 Expressing the right answers in spoken English
4 Using the correct basic English vocabulary in science writing
5 Solving problems in spoken English
6 Understanding exam questions in English
7 Using the correct English technical terms in science writing.

This set of language challenges refers to students' language challenges when reading, presenting, and communicating through English without any subject matter. It is similar to the concept of Basic Interpersonal Communication Skills (BICS), i.e., skills that are needed to perform any social interaction in a language or content-subject classroom. This component was labelled as 'challenges (use of basic English)'.

Table 7.3 Most heavily weighted factors for language challenges when learning science through English

Components	Descriptions	
Challenge (use of scientific English)	Language challenges related to comprehension, communicating in science content	Non-parametric
Challenge (use of basic English)	Language challenges related to the use of basic English	Non-parametric

In summary, the principal component analysis found two major components underlying the questionnaire data in students' responses to their language challenges. The two components were labelled and are shown in Table 7.3, with the most heavily weighted component in the top row of the table.

A Mann–Whitney test was conducted to determine whether there was a group difference between the early–full and late–partial EMI schools in regard to students' language challenges. The mean score of students' responses in the late–partial EMI schools in all the items was slightly higher than those of students' responses in the early–full EMI schools. This indicates that students in the late–partial EMI schools rated higher the same set of language challenges for learning science in English.

7.2.1   *The choice of instructional language in different science teaching activities*

There was a group difference in the choice of instructional language in a wide range of science teaching activities involving the students' science teachers. Three items were found to be relatively less significantly different, indicating that both early–full and late–partial EMI schools had less difference in the choice of instructional language for: discussing classwork with classmates; explaining the definition of scientific vocabulary; and explaining a science phenomenon. In addition, the mean score of the students' responses in the late–partial EMI schools has a higher value than that in the early–full EMI schools. This suggests that students in the late–partial EMI schools chose to have more code-switching between English and Cantonese and used more Cantonese for the mentioned science teaching activities.

7.2.2   *The most demanding English language skills for students to study science in English medium*

There was a significant difference between the early–full and late–partial EMI schools in terms of students' responses to the most demanding English language skills in the following five related English language skills: 'dialogue speaking skills'; 'writing daily assignments'; 'reading science textbooks'; 'listening to science teachers'; 'oral explanations in English', and 'distinguishing between English vocabulary for explanation and technical terminology'. As indicated by the five-point scale (1 = least demanding, 5 = most demanding), the mean scores of students' responses from the late–partial EMI schools in all the 17 items had a slightly larger value than those in early–full EMI schools. This echoes the findings from the scale on demanding English skills, for which students in the late–partial EMI schools rated English more demanding for the same set of language skills.

7.2.3   The most important English language skills

There was a significant difference between the early–full and late–partial EMI schools in students' responses to the most important English language skills in the following three related English language skills: 'Writing (e.g. daily assignments)'; 'Listening (e.g. science teachers' oral explanations in English)'; 'Expressing the right answers in simple English'. As indicated by the five-point scale (1 = least important, 5 = most important), the mean score of the students' responses in the late–partial EMI schools to all 17 items was slightly lower than those in the early–full EMI schools. This indicates that students in the late–partial EMI schools rated the level of importance lower than the early–full students for the same set of English language skills.

7.2.4   Students' self-concept of their science learning

There was no statistically significant difference between the early–full and late–partial EMI schools in terms of students' learning experiences in EMI classrooms in the students' responses to the overall scores. However, for the three items in Table 7.4, there was a significant difference in three out of four aspects

Table 7.4 The students' responses to their learning experience in EMI classrooms

Scale 4: Reflection. MOI	t test p value	Mean (S.D.)		
		Early–full	Late–partial	Both group
1. Before studying at my secondary school, I felt confident that my level of English would meet the demand of using EMI to study science subjects	0.078*	3.32 (0.92)	3.18 (0.94)	3.25 (0.93)
2. I felt I was at disadvantage to study science in EMI during junior levels	0.014*	2.76 (1.07)	2.98 (1.09)	2.87 (1.08)
3. I now feel confident my level of English meets the demands of using EMI to study science	0.006*	3.51 (0.83)	3.30 (0.96)	3.41 (0.90)
Overall score of students' reflection on their learning experience in EMI classrooms	0.320	3.20 (0.48)	3.15 (0.49)	3.18 (0.49)

Note: 1. The students' reflection on their learning experience in EMI classrooms was rated on a 5-point scale, where 1 = untrue of me, 2 = somewhat untrue of me, 3 = neutral, 4 = somewhat true of me, 5 = very true of me. 2.*$p < 0.05$; **$p < 0.01$; ***$p < 0.001$.

of reflection. Three differences emerge from the five-point scale (1 = untrue of me, 5 = very true of me). First, the mean score of students' responses in the early–full EMI schools was slightly higher for Statement 1, suggesting that these students were confident that their level of English would meet the demands of using English to study science subjects before they had started studying at secondary schools.

Second, the mean score of the students in late–partial EMI schools was slightly higher for Statement 2, suggesting that these students felt that they were at a disadvantage to study science in English during their junior levels. Third, there was a statistically significant difference between the two school groups. The mean score of the students in the early–full EMI schools was slightly higher than the students in the late–partial EMI schools. Statement 3 likely characterises the students in the early–full EMI schools more than the students in the late–partial EMI schools, suggesting that students in the early–full EMI schools felt confident that their levels of English would meet the demands of EMI.

7.2.5   *Students' belief in science classrooms*

There was no statistically significant difference between the early–full and late–partial EMI schools in terms of students' responses to the students' belief in science classrooms in students' responses in the overall scores. However, there was only one significant difference in students' belief when asking teachers questions related to the subject. As indicated by the five-point scale (1 = untrue of me, 5 = very true of me), the mean scores of the students' responses in the early–full EMI schools to most of the survey items (including students' beliefs when asking teachers questions related to the subject) were slightly higher than those in late–partial EMI schools. This suggests that students in the early–full EMI schools related more strongly to most of the statements. Students in the late–partial EMI schools rated lower the statements pertaining to lab experiments, suggesting that they may have more reflections on lab experiments.

7.2.6   *Students' views on the frequency of their use of coping strategies*

There were no statistically significant difference between the early–full and late–partial EMI schools in terms of students' responses to the frequency of using coping strategies. Three out of the nine items were found to have statistically significant differences. These items are: (1) checking Chinese translation of technical words; (2) understanding from examples; and (3) requesting more explanations from teachers in English. As indicated by the five-point scale (1 = untrue of me, 5 = very true of me), the mean score of the students' responses in the late–partial EMI schools to Item 1 had a slightly larger value, suggesting that students in the late–partial EMI schools used more Chinese translations of technical vocabulary. For Item 2, the mean score of early–partial EMI schools was higher, indicating that these students tended to understand the science concepts from more examples and/or rephrasing but that they did not need to translate them into Chinese. For Item 5, the mean score of early–partial EMI schools was higher, suggesting that these students might request additional English explanations from their science teachers.

7.2.7 *Students' views on the usefulness of their coping strategies*

There was a statistically significant difference between the early–full and late–partial EMI schools in terms of students' responses to the usefulness of their coping strategies. Six out of the nine items were found to have statistically significant differences. These items are Item 2 'checking Chinese translation of abstract English words'; Item 3 'understanding from examples'; Item 5 'requesting teacher's explanations in English'; Item 6 'requesting teacher's explanations in Chinese'; Item 7 'attending tutorials'; and Item 8 'requesting bilingual materials'. As indicated by the five-point scale (1 = untrue of me, 5 = very true of me), the mean scores of the students' responses from the early–full EMI schools for Items 2, 3, 5, 6, and 7 were slightly higher, suggesting that students in the early–full EMI schools found the following coping strategies more useful than students in the late–partial EMI schools: the Chinese translation of technical vocabulary; having examples in Chinese; requesting additional teacher explanations in both English and Chinese; attending tutorials; and requesting bilingual materials.

Drawing on the data from the survey regarding secondary school students' and teachers' perceptions on EMI, several conclusions can be drawn regarding the importance of such policy in both early–full and late–partial EMI schools. The results suggest that there are three possible groups of students in the early–full and late–partial EMI schools. The first group are students who prefer more Cantonese for interaction, find English language skills more demanding, and face more language challenges in science classrooms. These students frequently use coping strategies. While this profile might point to the poorly performing students in either type of schools, the questionnaire analysis showed that students in the late–partial EMI schools used coping strategies less frequently than those in the early–full EMI schools.

The second group is made up of students who believe that English language skills are important in EMI science classrooms, confident in their science abilities. These students also tend to prefer more English in classroom interactions and find coping strategies more useful. This group is likely to contain well-performing students in both types of schools. The third group consists of students who face many language challenges in the science classroom, use coping strategies frequently, and find these strategies useful. This group is likely to include students who are struggling with English but are motivated to overcome their language challenges.

7.3 Summary

This chapter has illustrated students' reflections on EMI in general, and the language challenges that students with various levels of English proficiency experienced in EMI science classrooms. Further, it has demonstrated students' views on the perceived usefulness and frequency of their use of coping strategies, self-concept in science learning, and belief in science classrooms. Taken together, it highlights the importance for schools to adopt and advocate a variety of appropriate coping strategies according to their students' individual language needs so as to facilitate their learning in science classrooms. In the next chapter, I will continue to explore the new factored variables included in the findings of the reported study, and propose appropriate coping strategies for students.

Reference

Gilanyi, L., Gao, X. A., & Wang, S. (2023). EMI and CLIL in Asian schools: A scoping review of empirical research between 2015 and 2022. *Heliyon*, *9*(6), 1–12.

8 Science students' perception of their EMI learning journeys

Findings on the factored variables

Among these 17 factors, there was a statistically significant difference in students' responses between early–full and late–partial English medium instruction (EMI) schools in the following nine factored variables (Table 8.1):

1a) Language choice for interactions (moderate effect size)
1b) Language choice for explaining science (small effect size)
2a) Demanding English (daily uses) (small effect size)
3b) Important English (basic) (relatively small effect size)
5a) Self-concept (science) (relatively small effect size)
7a) Challenges (use of scientific English) (small effect size)
7b) Challenges (use of basic English) (small effect size)
9a) Useful strategies (comprehension and science concepts) (moderate effect size)
9b) Useful strategies (learning in science) (small effect size)

For grade comparisons, there was a statistical difference only between Grades 10 and 11 in the early–full EMI schools for:

1a) Language choice for interactions (relatively small effect size)
8b) Frequency of using strategies (self-initiatives) (small effect size)
9a) Useful strategies (comprehension and science concepts) (small effect size)

There was also a statistical difference between Grades 10 and 11 in the late–partial EMI schools for:

2a) Demanding English (daily use) (small effect size)
2c) Demanding English (vocabulary) (small effect size)

DOI: 10.4324/9781003001454-9

Table 8.1 The 'factored' variables of each questionnaire scale and their normality

Scale		Major factors	Descriptions	Kolmogorov-Smirnov Z (sig)	Normality
1. Choice of instructional language in different science teaching activities	1a	Language choice for interactions	Language choice for interaction	**$p < 0.0001$*****	Non-parametric
	1b	Language choice for explaining science	Language choice for explaining science	0.011*	Non-parametric
2. The most demanding English language skills for students to study science in English medium	2a	Demanding Eng. (daily uses)	Demanding English skills related to daily uses	**$p < 0.0001$*****	Non-parametric
	2b	Demanding Eng. (scientific writing)	Demanding English skills related to scientific writing	**$p < 0.0001$*****	Non-parametric
	2c	Demanding Eng. (vocabulary)	Demanding English skills related to vocabulary	**$p < 0.0001$*****	Non-parametric
3. The most important English language skills for students to study science in English medium	3a	Important English (academic)	Important English skills related to academic language	**0.006****	Non-parametric
	3b	Important English (basic)	Important English skills related to basic language	**0.007****	Non-parametric
4. Student's reflection on EMI	4	Students' reflection	Overall score of students' reflection on their learning experience in EMI classrooms	0.320	Parametric
5. Students' self-concept of science	5a	Self-concept (science)	Self-concept in science learning	0.357	Parametric
	5b	Self-concept (cognitive. demand)	Self-concept in relation to cognitive demand	0.361	Parametric

(*Continued*)

Table 8.1 (Continued)

Scale		Major factors	Descriptions	Kolmogorov-Smirnov Z (sig)	Normality
6. Students' belief in science classrooms	6a	Students' belief (classroom activities)	Students' belief related to classroom activities	*p* < 0.0001***	Non-parametric
	6b	Students' belief (science-learning activities)	Students' belief related to their participations in science-learning activities	*p* < 0.0001***	Non-parametric
7. Students' views on language challenge	7a	Challenge (use of scientific English)	Language challenges related to comprehension, communicating in science content	*p* < 0.0001***	Non-parametric
	7b	Challenge (use of basic English)	Language challenges related to basic English skills	*p* < 0.0001***	Non-parametric
8. Students' views towards the frequency of the coping strategies	8a	Frequency strategies (external help)	Frequency of coping strategies related to external help	*p* < 0.0001***	Non-parametric
	8b	Frequency strategies (self-initiatives)	Frequency of coping strategies related to self-initiatives	*p* < 0.0001***	Non-parametric
9. Students' views towards the usefulness of coping strategies	9a	Usefulness of strategies (comprehension. Science concept)	Usefulness of coping strategies related to comprehension of science concepts	*p* < 0.0001***	Non-parametric
	9b	Usefulness of strategies (learning in science)	Usefulness of coping strategies related to learning in science classrooms	*p* < 0.0001***	Non-parametric

Note: *p* < 0.05; **p* < 0.01; ***p* < 0.001.

8.1   Early–full EMI vs. late–partial EMI & Grade 10 vs. Grade 11

8.1.1   *Language choice for interactions*

For the language choices related to interaction (1a), there was a group difference between the early–full and late–partial EMI schools. Students in the late–partial EMI schools more often chose Cantonese as their language for interactions. In terms of the comparison between grades, there was a statistically significant difference only in the early–full EMI schools between Grades 10 and 11. There was no significant difference between Grades 10 and 11 in the late–partial EMI school group.

8.1.2   *Language choice for explaining science*

For the language choice related to explaining science (1b), there was a group difference between the early–full and late–partial EMI schools. The late–partial EMI students chose to use more Cantonese when explaining science. There was no significant difference between Grades 10 and 11 in both types of EMI schools.

8.1.3   *Demanding English (daily use)*

There was a group difference in terms of students' responses to the most demanding English skills for daily use between early–full and late–partial EMI schools. Students in the late–partial EMI schools found daily-use English skills more demanding than their counterparts in the early–full EMI schools. With regard to the grade comparison, no significant difference was found between Grades 10 and 11 in the early–full EMI schools. The only statistically significant difference was found in the late–partial EMI schools between Grades 10 and 11.

8.1.4   *Demanding English (vocabulary)*

There was a significant difference in the late–partial EMI schools between Grades 10 and 11. Students in Grade 10 found English skills related to vocabulary more demanding than their counterparts in Grade 11.

8.1.5   *Important English (basic)*

There was a group difference for students' responses to the most important English skills (basic) between early–full and late–partial EMI schools. Students in the late–partial EMI schools found basic English skills more difficult than their counterparts in the early–full EMI schools. There was no significant difference between Grades 10 and 11 in either type of school.

8.1.6   *Self-concept (science)*

There was a group difference for students' responses to their self-concept in science between early–full and late–partial EMI schools. Students in the late–partial EMI schools had more confidence in their science abilities than their counterparts in the early–full EMI schools. There was no significant difference between Grades 10 and 11 in either type of school.

8.1.7   Language challenges (use of scientific English)

Group differences were found for students' responses to the language challenges related to comprehension and communication in science content between early–full and late–partial EMI schools. Students in the late–partial EMI schools found the same set of language challenges more difficult than their counterparts in the early–full EMI schools. There was no significant difference between Grades 10 and 11 in either type of school.

8.1.8   Language challenges (use of basic English)

There was a group difference in students' responses to language challenges related to basic English skills between the early–full and late–partial EMI schools. Students in the late–partial EMI schools found the same set of language challenges more difficult than their counterparts in the early–full EMI schools. There was no significant difference between Grades 10 and 11 in either type of school.

8.1.9   Frequent strategies (self-initiatives)

There was no statistically significant difference between early–full and late–partial EMI schools in terms of the frequency of using coping strategies related to self-initiatives. However, there was a significant difference in the early–full EMI schools between Grades 10 and 11. Students in Grade 11 used self-initiated coping strategies more frequently than their counterparts in Grade 10.

8.1.10   Useful strategies (comprehension and science concepts)

There was a statistically significant difference in the two factors pertaining to students' views on the usefulness of coping strategies. First, a group difference was found for students' responses to the usefulness of coping strategies related to the comprehension of science concepts between the early–full and late–partial EMI schools. The average mean score was higher for the early–full schools, suggesting that these students found the same set of coping strategies (comprehension in science concepts) more useful than their counterparts in the late–partial EMI schools. There was a significant difference only between Grades 10 and 11 in the early–full EMI schools, where students with higher grades found the coping strategies more useful.

8.1.11   Useful strategies (learning in science)

The second factor revealed a group difference in terms of students' responses to the usefulness of coping strategies related to learning in science classrooms between early–full and late–partial EMI schools. The average mean score of students in the early–full EMI schools was higher, suggesting that they found the same set of coping strategies more useful than their counterparts in the late–partial EMI schools. There was only a significant difference between Grades 10 and 11 in the early–full EMI schools. Students in Grade 11 found the same set of coping strategies (learning in science) more useful than their counterparts in Grade 10. There was no significant difference between Grades 10 and 11 in both types of schools.

8.1.12   Summary of students' perceptions of their EMI experiences (the 17 factored variables)

The major findings of the students' perceptions of their EMI experiences are as follows:

1   Science students in the early–full and late–partial EMI schools had different perceptions on the following: the preferred language choice for interactions and for explaining science; which English skills are most demanding; which English (basic) skills are most important; language challenges; and the usefulness of coping strategies.
2   Students in the late–partial EMI schools preferred more Cantonese for interactions and for explaining science.
3   Students in the late–partial EMI schools believed that English language skills for daily use were more demanding.
4   Grade 11 students believed that English skills related to daily-use vocabulary were more demanding than did students in Grade 10.
5   Students in the late–partial EMI schools thought basic English language skills for daily use were less important than did students from the early–full schools.
6   More students in the late–partial EMI schools believed they could do well in science than those in the early–full EMI schools.
7   Students in the late–partial EMI schools reported that they had more language challenges in the areas of comprehension and communicating science content, as well as in basic English language skills, than did students in the early–full schools.
8   Students in the late–partial EMI schools reported that they used self-initiated coping strategies less frequently than did the students in the early–full EMI schools.
9   Students in the late–partial EMI schools reported that they found coping strategies to improve their comprehension of science concepts less useful than did the students in the early–full EMI schools.
10   Students in the late–partial EMI schools reported that they found coping strategies to enhance their learning in science less useful than did the students in the early–full EMI schools.

To conclude, the questionnaire results show that students in the late–partial EMI schools preferred more Cantonese for interactions and for explaining science.

Although these students believed that daily English language skills and basic vocabulary are more demanding, they found these skills less important than their peers in the early–full EMI schools. This was also observed among the late–partial EMI teachers (in Chapter 6). In addition, students in the late–partial schools believed that they had the ability to do well in science, but they encountered more language challenges related to comprehension and communicating in science content as well as in basic English skills. In terms of coping strategies, late–partial students used self-initiated coping strategies less frequently than students in the early–full schools. Late–partial students also found strategies used for enhancing their comprehension of science concepts and learning of science to be less useful than students in the early–full EMI schools. In the next chapter, I will summarise the analysis based on the data collected from the student focus groups.

8.2   Science student's views: analysing the students' focus group interviews

This section elaborates on the results of the semi-structured interviews conducted with the focus groups. It focuses on seven specific aspects of science students' perceptions, including:

1 their language choice for interactions as well as the barriers they encounter in science classes;
2 their use of vocabulary when studying science subjects;
3 the coping strategies they adopt in relation to vocabulary;
4 the language issues they face when writing and reading in science;
5 their coping strategies for writing and reading in science subjects;
6 how they seek external help;
7 their positive and negative views of EMI teaching.

8.2.1   *The background of the student interviewees*

Fifty-two science students from 33 lessons were recruited to participate in focus group interviews. I invited five to eight randomly selected students from each lesson to attend the interviews after the observations. Teachers were not asked to nominate students. If students could not attend the interviews, I arranged another interview on the same day of the observation. Students who participated in the interviews were invited to discuss the topic freely but I guided the discussion according to the interview protocol (discussed in Chapter 3). Students shared their opinions about the use of Cantonese in their EMI experiences, possible language challenges they faced when studying science through English, and the coping strategies they used. The interviews lasted from 20 to 40 minutes, and each of the students had at least 2–3 minutes to speak during the discussion. All students identified Cantonese as their first language. This largely reflects secondary school students' background in Hong Kong. Each of the participating students was given a pseudonym and a code (S1 to SX) to be referred to in the subsequent analysis. In addition, students were invited to share their experiences responding to teachers' questions or discussing with classmates in English, as well as their opinions on helpful question types for understanding science, the kind of vocabulary that is important for learning science, and their expectations of their English proficiency in learning science.

8.2.2   Results of the students' interviews

The interviews were analysed based on seven different areas related to students' perceptions of their EMI learning experiences:

1 language choice for interactions, and the challenges they face in the classroom;
2 vocabulary in science subjects;
3 the coping strategies related to vocabulary;
4 language challenges related to writing and reading in science;
5 coping strategies for writing and reading in science;
6 strategies related to external help;
7 positive and negative attitudes towards EMI.

8.2.3   Language choice for interactions and the challenges

The majority of the students in the early–full and late–partial EMI schools highly valued the role of L1 in learning science through English, particularly when interacting with their science teachers or working with their peers during a lab experiment. Students in the late–partial EMI schools strongly believed that their L1 was an easy and efficient means for communicating science in lab operations, as the following two examples show:

> It is because mother tongue is the most convenient language to express the message I want to say. If we use English, we would waste a lot of time to communicate and ignore the importance of the experiment.
>
> (Student 1, late–partial EMI)

> I will respond in mostly Cantonese, when I don't understand what the teacher said in English.
>
> (Student 3, late–partial EMI)

Students also reported that more Cantonese was spoken in biology and chemistry than in physics. For instance, Student 2 said:

> I use Chinese more in biology class as well as chemistry class. But I would use English if the situation is okay. And if it doesn't work, I would mix Chinese with English.
>
> (Student 2, late–partial EMI)

Despite the usefulness of Cantonese, the students highlighted a few possible consequences of using L1:

> When answering questions, I know the Chinese but forgot their English words. Like before … that's because he uses Chinese to memorise.
>
> (Student 3, late–partial EMI)

These students' comments resonate with other studies where students were found to rely on L1 during collaborative science discussion and practical work (Pun & Cheung, 2021). Students can utilise their L1 to connect with prior knowledge and everyday experiences, interpret science tasks, and construct science knowledge (An & Thomas, 2021). Too much of L2 use has also been suggested to hinder effective classroom interactions regarding meaning negotiation and scaffolding, leading to teacher-centred classrooms and poor learning outcomes (Lo & Macro, 2012). On the other hand, the last student's comments remind EMI teachers of the caveat of heavily exploiting students' L1 to circumvent potential comprehension or interaction challenges. In fact, many students also recognised the importance of English skills for their learning in science education (Pun, 2022).

8.2.4   *Coping strategies for overcoming the challenges of communicating in English*

Some students with less competence in English try to organise their thoughts in their mother-tongue (Cantonese) first and then translate them into English. Students called this way of translating ideas from Cantonese to English 'Chinglish'. The students felt that this is an efficient way for those with limited English skills to express science ideas. As Student 3 said:

> For me, I would use English to express myself if I can. If I can't, I would use Chinglish.
>
> (Student 3, late–partial EMI)

Others with poor English skills reported that they relied on body language in order to communicate with their teachers in English. For instance, Student 2 explained:

> When the teacher asks you a question, but I can't express it in English, I would try my best to use body language to help my teacher to understand me when I cannot express myself.
>
> (Student 2)

In addition, when students struggled with abstract science explanations given by their teachers, many of them waited for their teacher to 'recap' the key concepts in

L1 and then in English. Students would only use the 'simplest and condensed' L2 expressions. Here are two examples:

> In most situations, I would use Chinese to tell my teacher and wait for the English explanation of the teacher.
>
> (Student 2, late–partial EMI)

> 'Researcher: [What would you do] if you were asked to write the observable result from today's experiment?
> Student: I would use the simplest expression "This is zinc" rather than 'in this experiment, zinc was generated and we observed a metal layer was formed and that was zinc metal.'
>
> (Student 4, late–partial EMI)

These students, though with limited English language proficiency at that time, actively adopted self-regulatory learning strategies and capitalised on resources from their L1 knowledge, L2 knowledge, body language, and teachers to communicate in science classrooms in English. This may be explained by the performance-oriented characteristics of East Asian students found by Chen and Wong (2015). Their goals towards better academic achievement contribute to their adaptive learning and use of coping strategies in the face of copious challenges during learning in EMI classrooms (Pun, 2022).

8.2.5   *Vocabulary in science subjects and coping strategies*

8.2.5.1   *Vocabulary in science subjects*

Student interviewees in both early–full and late–partial EMI schools emphasised their difficulties in learning English vocabulary in science subjects. They found that there were too many specific science terms. For instance, in biology, a certain number of vocabulary regularly appear in a variety of related science topics. Students claimed that they had more problems remembering all the vocabulary, particularly the descriptive vocabulary:

> 'I think … this … bio … there is too much vocabulary.' (Student 1, late–partial EMI)
> 'So, it is very hard to memorise them all … Those kinds of vocabulary that describe bones and brain etc. Don't quite remember. I forgot them right after the exam.'
>
> (Student 3, late–partial EMI)

8.2.5.2   *Words with multiple meanings and the process of nominalisation*

Another challenge concerns scientific terms with multiple meanings in different science subjects. Students in the early–full EMI schools with high-level proficiency in English still found it puzzling that the same words have multiple meanings in science. For example, 'normal' in physics means an object such as a line or vector that is perpendicular to a given object, but in other science subjects like chemistry or biology, it suggests the meaning 'ordinary'.

Students in both early–full and late–partial EMI schools reported that sometimes it is very challenging to use scientific words to convey simple meanings because students are not familiar with the different grammatical or semantic categories of words such as synonyms, derivations, and verbs. One student reported that using action verbs to describe scientific processes was challenging. He found it very difficult to rephrase them into everyday language and to change the grammatical configuration of these words (e.g., neutralise—neutralisation), or what linguists call the process of 'nominalisation' (Halliday & Martin, 1993). The student said he needed to memorise the same words with different grammatical forms for different semantic contexts:

> Verbs to describe the process are more difficult. The solutions are … to read more and do more references.… I can memorise those terms if I really study hard, but it's hard to change the form of a word and to come up with the verb.
>
> (extracted from Student 2, late–partial EMI)

To date, challenges related to disciplinary-specific language, especially in dealing with science language, have been well documented in literature (e.g., Evans & Morrison, 2011; Lee et al., 2020; Lewis et al., 2011; Yip et al., 2003). Similar to what has been mentioned in the above-mentioned two students' excerpts, the challenges centred on the diverse technical vocabulary, specialised lexico-grammatical features, and grammatical metaphors in the language of science.

8.2.5.3   *Coping strategies related to vocabulary*

Students in both early–full and late–partial EMI schools suggested that there are a number of coping strategies useful for overcoming the challenges related to vocabulary in science. These accounts display individual differences in L2 vocabulary learning (Li et al., 2022). The first is rote memorisation. Students in early–full EMI schools worked hard at spelling, understanding the meaning in the context, and word association. They made notes with a list of examples in complete sentences. This helped them develop their vocabulary by connecting their pronunciation and word meanings:

> I guess the most difficult part is those new terms when using English to learn. For example, in physics, chemistry and biology, there are some words that have a different meaning from the meaning we usually perceive. So, I think the solution is to spell more and do more exercises.
>
> (Student 2, late–partial EMI)

Sometimes we need to use a more complicated word to express a simple meaning. And my solution is jotting down those words if the teacher has said them. Or when I go home, I may go to check the meaning of the sentences if they are really difficult.

(Student 3, late–partial EMI)

The students in both types of EMI schools stressed that memorising vocabulary meant that they had truly to understand both the English and Chinese meanings. For students with poor English, knowing the Chinese meaning helped them in memorisation. However, some students believed that memorising did not help them truly understand word meanings. Instead, they argued that one should understand how to use the word according to its context and make a connection to its scientific meanings.

Students also reported that a second strategy was to express the same meanings by using synonyms (e.g., affect, influence) in order to increase their vocabulary in science:

As for writing the observation, Mr. Kwok (teacher) reminded us that we should not use the word 'affect' if we are trying to convey the meaning 'influence' in experiments. We need to use another word to substitute it. Also, using complete sentences to write is difficult as well. These are the difficulties I think. And the solution is the same, to revise the content in the textbook and write according to the structure Mr. Kwok taught us before.

(Student 4, late–partial EMI)

A third strategy mentioned by the students was to try to understand the word-associated meanings in similar contexts. For example, the word 'dissolve' in chemistry can have different meanings according to the context. When it refers to a substitution reaction, it is related to the process of dissociation, meaning a substance is dissolving into a solution. When it refers to an electrochemical cell, it is related to the process of ionisation, meaning the ions are dissolving into an electrode.

For example, you don't know the Chinese meaning of the terms 'dissolve'. I memorise the English term ... for example ... I memorise 'dissolve', I know it is about dissociation, when it is related with substation reaction, but when it is related to electrochemical cell, it is about ionisation ... The most important part is to describe the word-associated meaning.

(Student 5, early–full EMI)

8.2.6 *Language challenges related to writing and reading in science*

When dealing with the demand of writing and reading in science, students in the late–partial EMI schools in particular found it hard to write a good explanation with the correct choice of key scientific terms. Students complained that they could not locate the key concepts in writing science explanations in an exam. Some students relied on their thinking in their first language in order to convey a complete explanation but found that it did not match with what they expressed in English. In other words, there was a discrepancy between what students understand in Cantonese and what they produce in English. Students complained that they did not know the 'accurate terms' to use in science explanation. This issue may be particularly evident in the late–partial EMI schools where students produce their explanations by rote memorisation. Students also seemed to have difficulty in picking out relevant key scientific words that were important to remember if their teachers or textbooks did not highlight the scientific words. Instead, they wrote down every single English sentence from their science textbooks in order to solve all related exam questions without truly understanding them. For example, an early–full student described about his challenges in reading and writing science:

> I always cannot write the key points. That is not getting the main points ... I don't know why I couldn't get the mark ... I missed some details, and maybe then I couldn't get the marks.
>
> (Student 2, early–full).

Another early–full EMI student described her problem with accurate terms. She said:

> Incomplete meaning? It was indeed quite complete. Like, it was following my way of thinking. Like I had this thought, but maybe I didn't use the accurate terms ... Maybe I saw it in the book, and my teacher didn't really talk about the terms. And then we answered at last.
>
> (Student 2, early–full)

Students also reported other challenges related to reading and writing in science. First, students struggled with question words, or not knowing the nature of questions. For example, with question words like 'account, analyse, exemplify', some students would misinterpret these words as asking them to show the calculation rather than to produce a narrative to describe and explain their reasoning:

> Sometimes when you look at the questions, there are some words you don't know ... e.g. solving the question.
>
> (Student 1)

Second, students struggled with unfamiliar terms in a topic they were studying, for example, the name of a disease, bacteria, or a chemical compound. Students may learn the word 'acetylsalicylic acid' from their textbooks but do not know it is called 'aspirin' in daily life as they have only heard the Chinese name from

TV (阿士匹靈, direct translation from English 'a+s+pr+in'). As one student explained:

> They are technical terms. Like some diseases. Like the one... Glucose is OK. But hypoglycaemia, this term is included in the textbook. Since we have learnt it before and we know what it means, we can guess the question. But sometimes, there are questions about some symptoms that we don't know. But you are required to write down the functions of the diseases, or their harm, or how the diseases are formed.
>
> (Student 2, late–partial)

Reading and writing (e.g., summarising previous lessons, reading in journal clubs) are two of the most frequent activities in secondary schools, and EMI contexts are no exceptions (Pun et al., 2023). The challenges faced by students in this study, resonating with other scholars who also pointed out the challenges for EMI science students in writing science-related materials and comprehending complex sentences (An & Thomas, 2021; Lu et al., 2021), further accentuate the urgency of catering to their receptive English language skills in science learning.

8.2.7 Coping strategies for writing and reading in science

When students discussed the coping strategies for writing and reading in science, the majority of them in both types of schools stressed the need for explicit teaching about the structures and sentence patterns expected in their science writing. One student from the early–full EMI school with high-level English proficiency still stressed that she needed to read reference books on the subject and do more practical exercises in order to better understand the structure and key points needed to succeed at writing about science. She suggested that science teachers should focus on key concepts rather than spend lots of time on elaborations or explaining daily examples:

> Although writing complete sentences and paragraphs might be more difficult for me, teachers sometimes teach us how to write to build a better structure and to get the points. As for my solution, I would read more reference books on the subject and do some practice exercises before exams and during practice. [...] You (the teacher) can keep repeating the points but not say lots of other things and leave the main point to the very end and point out the main point directly and give some examples as we won't understand the main point if there are just a few examples.
>
> (Student 4, late–partial)

The majority of other strategies relied on external help from tutorial centres. Students valued the role of tutorial centres where they could reinforce their understanding and highlight the key concepts through dedicated L1 instruction. For instance, Student 2 reported:

> I go to chemistry tutorial classes. I think the tutorial class teaches too fast. I am in form 4 now but it already taught the curriculum after form 5 so I quit it. But what the class teaches is rather common. It will teach the content in the textbook more deeply and help you understand more. Also, it uses Chinese in teaching which makes it easier to understand.
>
> (Student 2, late–partial EMI)

Having out-of-school tutorial classes is not only the case in Hong Kong. Studies in other Asian contexts also found tutorials as one way to help EMI students better comprehend content knowledge and promote their English learning (Qiu & Fang, 2022; Ruegg, 2021). Although this seems to be an effective way of coping with reading and writing challenges, it also prompts us to ask how to improve EMI teaching quality to better support students so that they do not have to spend extra time and economic resources. One possible way, as can be implied from the student's excerpts, is to listen to students' voices to discover their learning demands and expectations from teachers. In this way, students can improve their motivation in learning since teachers are responsive to their needs; EMI teachers can also be empowered to make decisions on what to focus (i.e., language or content knowledge) in a specific science task.

8.2.8   *Positive and negative attitudes towards EMI*

Students in both early–full and late–partial EMI schools had mixed feelings towards their EMI experiences. While many students in the late–partial EMI schools saw EMI as an opportunity to improve their English, their counterparts in the early–full EMI schools believed that learning science through English had perhaps caused them to be less motivated to learn both science and English. One student from a late–partial school explained:

> And I think using English to teach science subjects is quite good because my English is not bad, so spelling is not a problem for me.
>
> (Student 4, late–partial EMI)

On the other hand, one early–full student said:

> I think I am less interested to learn science in English compared with learning science in Chinese. The difficulties are the words are difficult to understand and to use such as some technical verbs.
>
> (Student 2, early–full EMI)

A few students in the early–full EMI schools welcomed more L1 in their EMI science classrooms to help them understand abstract concepts and communicate more effectively with teachers and their peers. They thought that L1 was better than L2 for learning science. There may be multiple reasons that account for the students' different attitudes towards EMI. In late–partial EMI schools, these results resonate with other studies (Aguilar & Rodríguez, 2012; Tatzl, 2011), in which they found that students believe that learning content courses through English could benefit their English language proficiency. Their positive attitude may also be driven by the status of English as a useful medium of communication in daily and academic life in HK, a multilingual and multicultural society. In contrast, students in early–full EMI schools implied that when examining from the content knowledge acquisition perspective, it is doubtful for many students, not only in this study's case, that EMI hinders their comprehension of subject knowledge to some extent (Byun et al., 2011; Costa, 2017).

8.2.9 Summary of the interviews: students' perceptions of EMI learning in science classrooms

The above analysis explored early–full and late–partial students' views on science learning in EMI classrooms. The students in both schools shared similar views in many aspects. First, one important finding is that both early–full and late–partial EMI students highly valued the role of L1 in learning science through English, particularly when interacting with their science teachers or working with their peers during a lab experiment. Students with limited English in both EMI groups would use body language or 'Chinglish' to help them communicate effectively. Some of these students would rely on their science teachers to 'recap' the key concepts in L1 and then in English.

Both early–full and late–partial EMI students commented on the challenges of learning vocabulary in science subjects. They struggled with the amount of specific science terms, words with multiple meanings, connotations, and word-associated meanings. Some students found it difficult to use scientific words to convey simple meanings because they did not know how to manipulate the different grammatical forms and semantic categories. Students in both groups suggested a number of ways to overcome these challenges, including rote memorisation, using synonyms, understanding the word-associated meanings in science contexts, and making connections with a word's function in science and in daily life.

Students in both groups pointed out a number of language challenges related to writing and reading in science, but there were bigger and more challenges for the students in the late–partial EMI schools with limited English abilities and experiences. Students found it very challenging to convey a complete meaning in their science explanations, and they sometimes found that what they wrote in their examinations did not match with what they could express in spoken Cantonese. Students complained that they did not know the 'accurate terms' to use in science explanation. In addition, students reported that they struggled to identify relevant key scientific ideas, guess the meaning of an unfamiliar word in a topic (e.g., a disease's name), and understand the nature of examination question words.

Students in both groups suggested a number of strategies. They stressed the need for explicit teaching about the structures and sentence patterns expected in their science writing. They also sought help from tutorial centres, where they received more guided support while working on exercises based on past public examination papers on a range of science topics. The tutorial centres also provided support in essay writing in a well-organised manner. The only drawback, according to some students, was that these tutorial centres were exam-oriented.

8.3 Summary

This chapter has reported on the interview data pertaining to students' perceptions of their EMI experiences in both early–full and late–partial EMI schools. Seven recurring themes were identified from the focus group students' interview transcripts. These themes, as discussed above, are (a) language choice for interactions, and the language challenges they face in the classroom; (b) vocabulary in science subjects; (c) the coping strategies related to vocabulary; (d) language challenges related to writing and reading in science; (e) coping strategies for writing and reading in science; (f) strategies related to external help; and (g) positive and negative attitudes towards EMI. In the next chapter, I will summarise all the major findings and triangulate data from multiple sources, including classroom observation, survey and interview data, in order to answer each of the research questions.

References

Aguilar, M., & Rodríguez, R. (2012). Lecturer and student perceptions on CLIL at a Spanish university. *International Journal of Bilingual Education and Bilingualism, 15*(2), 183–197.

An, J., & Thomas, N. (2021). Students' beliefs about the role of interaction for science learning and language learning in EMI science classes: Evidence from high schools in China. *Linguistics and Education, 65*(1), e100972.

Byun, K., Chu, H., Kim, M., Park, I., Kim, S., & Jung, J. (2011). English-medium teaching in Korean higher education: Policy debates and reality. *Higher Education, 62*(4), 431–449.

Chen, W. W., & Wong, Y. L. (2015). Chinese Mindset: Theories of intelligence, goal orientation and academic achievement in Hong Kong students. *Educational Psychology, 35*(6), 714–725.

Costa, F. (2017). The introduction of English as an academic language in a faculty of physics and mathematics in Italy. *L'Analisi Linguistica e Letteraria, 25*(2), 269–288.

Evans, S., & Morrison, B. (2011) Meeting the challenges of English-medium higher education: The first-year experience in Hong Kong. *English for Specific Purposes, 30*(3), 198–208.

Halliday, M. A. K., & Martin, J. R. (1993). *Writing science literacy and discursive power.* London: Falmer Press.

Lee, E. N., Orgill, M., & Kardash, C. (2020). Supporting English language learners in college science. Classrooms insights from chemistry students. *Multicultural Education, 27*, 25–32.

Lewis, S., Maerten-Rivera, J., Adamson, K., & Lee, O. (2011). Urban third grade lecturers' practices and perceptions in science instruction with English language learners. *School Science and Mathematics, 111*(4), 156–163.

Lo, Y. Y., & Macro, E. (2012). The medium of instruction and classroom interaction: Evidence. From. Hong Kong secondary schools. *International Journal of Bilingual Education and Bilingualism, 15*(1), 29–52.

Lu, C., So, W., Lee, Y., & Yeung, Y. (2021). Exploring the problems of learning science in the English medium: A study on high school students' perceptions and attitudes in China. *Asia Pacific Journal of Education, 43*(4), 1–16.

Pun, J. (2022). Effects of prior English exposure on Hong Kong tertiary science students'. Experiences in EMI learning. *RELC Journal*, e3368822210790.

Pun, J. K., & Cheung, K. K. C. (2021). Meaning making in collaborative practical work: A case study of multimodal challenges in a year 10 chemistry classroom. (2021). *Research in Science & Technological Education, 41*(1), 271–288.

Pun, J. K. H., Fu, X., & Cheung, K. K. C. (2023). Language challenges and coping strategies in English Medium Instruction (EMI) science classrooms: A critical review of literature. *Studies in Science Education*, 1–32.

Qiu, X., & Fang, C. (2022). Creating an effective English-Medium Instruction (EMI). Classroom: Chinese undergraduate students' perceptions of native and non-native English-speaking content teachers and their experiences. *International Journal of Bilingual Education and Bilingualism, 25*(2), 641–655.

Ruegg, R. (2021). *Supporting EMI students outside of the classroom: Evidence from Japan.* Routledge.

Tatzl, D. (2011). English-medium masters' programmes at an Austrian university of. Applied sciences: Attitudes, experiences and challenges. *Journal of English for Academic Purposes, 10*(4), 252–270.

Yip, D. Y., Tsang, W. K., & Cheung, S. P. (2003). Evaluation of the effects of medium of instruction on the science learning of Hong Kong secondary students: Performance. On the science achievement test. *Bilingual Research Journal, 27*(2), 295–331.

9   Closing the gap between policies and practices

EMI research in theory and practice

9.1   Summary of overall findings

This study explored classroom interactions in different EMI classrooms and the science teachers' and students' attitudes towards their EMI experience. It did not seek to evaluate the schools, teachers, or students, and the findings are by no means an indicator of the quality or effectiveness of different types of EMI schools. It adopted a research design from Lo and Macaro (2012) to extend the scope of investigation on classroom interaction to the previously unexamined context of senior secondary science classrooms in Hong Kong. Nineteen teachers and 545 students from Grades 10 and 11 EMI science classrooms were recruited from 8 schools: 4 'early–full EMI' (full English medium of instruction from Grades 7 to 12) and 4 'late–partial EMI' (Chinese medium of instruction for Grades 7–9, and then switching to partial EMI instruction for Grades 10–12). It used multiple sources of data collected from video-recorded classroom observations and questionnaires and interviews with teachers and students. All data were analysed quantitatively and qualitatively, considering the choices of instructional language for teaching science, teachers' beliefs about the effectiveness of EMI, students' self-concepts in science, the language challenges that students encountered, and the coping strategies that students used when learning science through English. In the following sections, I summarise the key findings and describe how they answer the four research questions of this study.

9.1.1   Research Question 1: What is the nature of classroom interaction in EMI science classrooms?

Despite the schools claiming that they were EMI schools, there was diversity in the actual amount of English used between the early–full and late–partial EMI schools. It is quite evident that the four early–full EMI schools mostly used English as the medium of instruction, while two of the four late–partial EMI schools used more Cantonese than English as the instructional language.

However, regardless of the MOI, the nature of interaction patterns in both early–full EMI and late–partial EMI schools was found to be similar in this study. I could not find any examples of negotiation of meaning or L2 learning, particularly of

DOI: 10.4324/9781003001454-10

students making clarification requests. The nature of interaction was short, usually through the simple Initiation–Response–Feedback (IRF) sequences, which tended to limit opportunities for teacher–student negotiation of meaning and understanding. This echoes a number of studies that have compared the interaction between early–full EMI and late–partial EMI schools in Hong Kong. Lo and Macaro (2012) observed that when the MOI was changed from Cantonese to English, the teacher–student interaction became very minimal and took the form of short interactive sequences. Yip et al. (2007a, 2007b) also found the quality of instructional activities in these EMI science classrooms to become very limited, with little interaction between teacher and students and the increased use of recall-type questions. Lin (2006) observed interactive teaching practices in some mixed-mode science classrooms in which teachers delivered the content in L1 and recapped it in L2. Whereas these previous studies suggest that late–partial EMI schools are more interactive, and are therefore different from early–full EMI schools, the current study reveals that lessons from both types of EMI contexts tend to be authoritative and less interactive.

The instructional approach of classroom interactions in the lessons observed was authoritative, very often with teachers dominating interactions in an effort to relay their specific viewpoint on a science topic. This is evident in the qualitative analysis of transcripts from the four sampled biology EMI science classrooms. Interactions between teachers and students were limited, with very few extended IRF patterns. The nature of teacher–student interaction changed, with fewer cycles of prolonged I-R-F-R-F sequences.

Most late–partial EMI schools did not have any opportunities for teachers to negotiate the meaning of words with students or provide 'scaffolding' for concepts and processes in the academic curriculum. This echoes Lo and Macaro's (2012) observation of late–partial EMI (MOI-switching) schools, which tended to be more teacher-centred in which students' participation was inhibited. The interaction patterns in both types of EMI schools were reduced from Grades 10 to 11, and extended IRFRF sequences were rarely found in Grade 11.

In the current study, students seldom initiated interactions and were mainly invited by teachers to give responses that showed their understanding of the science topics. Teachers tended to select which students' responses were responded to, followed up, or ignored. The teachers mainly checked students' understanding, shared their own ideas, and highlighted key ideas that matched their planned science views. This is again consistent with Lo and Macaro's observation of the MOI-switching teachers who appeared less skilful in helping students answer cognitively demanding questions.

The authoritative nature of teaching suggests that the teachers perhaps lacked instructional strategies for interacting with students in L2. This reflects the findings of previous studies which have revealed very few instances of negotiation of meaning at the lexical level among content-subject teachers in EMI classrooms (Lo & Macaro, 2012). The teachers could have encouraged students to arrive at conceptual understandings themselves through higher-order pedagogical skills (e.g., clarification requests, confirmation checks and signals, comprehension checks) and match their existing knowledge in L1, thereby promoting their students' deeper understanding of the science concepts and processes. Instead, the EMI teachers in this study tended to adopt lower-order pedagogical scaffolding, by only 'informing' their students of the facts. This strategy subsequently encouraged students' rote memorisation of factual information, a finding in line with the study by Yip et al. (2007), in which many EMI science classrooms were found to promote rote memorisation of facts.

This study found that language use, in terms of developing students' understanding of science knowledge, was similar between the early–full and late–partial EMI schools. This study also found that the science language used by teachers in classroom interaction was high in the use of: (1) subject–specific vocabulary; (2) nominalisations and grammatical metaphors; (3) complex noun phrases; (4) language of knowledge relationships; and (5) lexical density. These observations are similar to Kong and Hoare's (2012) findings that teachers in junior secondary science classrooms in Hong Kong used complex academic language (i.e., a high proportion of subject–specific vocabulary and nominalisation, complex noun phrases, and high lexical density in teacher's spoken language use).

In term of students' language production, in the early–full EMI schools, students did not appear to be able to interact with teachers as freely and fluently as their counterparts in the late–partial EMI schools. Students in the early–full EMI schools had experienced EMI for more than three years and might have become accustomed to EMI teaching. This finding contradicts previous studies such as Marsh et al.'s (2000) six-year longitudinal study, which showed that early–full EMI students would 'catch up' in their English in senior forms, and therefore the disadvantages of using EMI in earlier years would have been offset at the later stage of secondary education. This finding is also counter to Cummins' (1979) proposal that it may take three to five years for students to reach the threshold level required to benefit from immersion programmes. The disadvantage of using EMI is not simply a time issue but rather necessitates the provision of conditions such as more teacher input modification in L2, more teacher allocated time for whole-class interaction, and the availability of bridging courses to improve students' English proficiency, which could enhance the learning outcomes of all students, especially for those with lower language proficiency (Chan, 2014).

The qualitative analysis of lesson transcripts in Chapter 5 revealed that students in both early–full EMI and late–partial EMI schools predominately produced a number of incomplete sentences consisting of short, technical nouns or noun phrases referring to scientific items. Science teachers rarely made any attempts to correct their students' language mistakes or to encourage students to produce a complete sentence. This lack of teacher feedback on students' L2 language production reflects the fact that EMI science teachers rarely provide any input to facilitate students' L2 language learning.

As a result of their language shortfalls, the students' L2 remains underdeveloped despite instruction in English. This reality of a lack of language support perhaps reflects a gap between the aim of the fine-tuning EMI policy and its actual implementation. The original aim of the fine-tuning policy was to create an additional English environment to increase students' exposure to English in order to improve their English proficiency (Poon, 2013). The lack of students' language output in English in Hong Kong EMI classrooms appears similar to that of Canadian immersion programmes which found that students fell short of advanced-level language skills, especially in oral production. Swain (1995) also pointed out the important role of 'pushed output' in achieving effective immersion programmes.

This study did not set out to test students' comprehension of language or content. Nonetheless, based on its findings, it would seem that, in the four biology EMI classrooms in both the early–full and late–partial EMI schools, the adoption of EMI perhaps only led to the development of students' comprehension of content knowledge, rather than the development of their proficiency in language production. The absence of support and pushed output on students' language production from their teachers is perhaps one of the major challenges in achieving the goal of improving EMI students' English skills, both receptive and productive.

The lack of language support for students and pedagogical support for teachers in EMI science classrooms found in this study can perhaps explain why Hong Kong senior secondary students experience tremendous language challenges during their first year at university (Lin & Morrison, 2010). As pointed out by Lin and Morrison (2010), the adoption of CMI teaching had a significant negative effect on the size of the academic vocabulary of Hong Kong tertiary students. Thus, students in CMI schools were disadvantaged during their first year of university study. Most teachers had to use a mixture of L1 and L2 in order to give clear instructions to students regarding different science concepts, reflecting the fact that insufficient support was given to teachers to assist them in the change of MOI policy at schools; hence the teachers found that using solely English as their MOI would be inefficient in the classroom. Given the results of the present study, we can predict similar results from students in the late–partial EMI schools by comparing different perceptions of teachers and students regarding their teaching and learning experiences. This again contradicts the findings of Marsh and his colleagues (2000) in a six-year longitudinal study on the impact of MOI on secondary school students' English and other content-subject achievements. In that study, early–full EMI students were found to 'catch up' in their English proficiency by senior forms, and thus the disadvantages of using EMI were offset at a later stage of secondary education.

The analysis of the four biology lessons echoed Lin and Wu's (2015) argument of the role that teacher-modified input plays in creating space for 'translanguaging'. They reported on a junior biology secondary science teacher in Hong Kong who skilfully allowed a low-proficiency student to contribute to the class discussion in L1 and guided the whole class to discover the equivalent meaning in L2. In my analysis, although the two EMI teachers spoke mostly in English, they skilfully intervened with their students' responses at the feedback stage by either expanding, acknowledging, or ignoring students' ideas in developing students' understanding of the scientific story. Teachers' modified input through extended IRF exchanges, either in L2-only or by translanguaging between L1 and L2, serves as an important way of providing comprehensive feedback to students in developing their scientific knowledge. The lesson examples in Chapter 5 from the early–full and late–partial EMI biology classrooms have reference value for other EMI teachers who are routinely faced with the challenge of teaching both content knowledge and language through L2.

9.1.2   *Research Question 2: What are the differences and similarities in classroom interaction between early–full and late–partial EMI schools?*

In Chapter 4, I showed that the interaction time and distribution of teacher and student talk are similar in early–full and late–partial EMI schools. Moreover, the percentages of interaction time between the early–full EMI (69%) and the late–partial EMI (73%) schools were similar. In both school types, 62% to 68% of the interactions were in teacher talk, with very little student talk (5–8%), and the average length of a teacher turn was longer than that of a student turn, suggesting that these EMI lessons were teacher-centred, authoritative, and very often with teachers dominating interactions in an effort to relay their specific viewpoint on a science topic. This is different from Lo and Macaro's (2012) research in which they found that the mean length of teacher and student turns increased significantly from Grade 9 to 10 in the MOI-switching schools.

In addition, a high percentage of L2 (93.5% vs L1: 6.5%) use was observed in the early–full EMI schools, and a high percentage of L1 (53% vs L2: 44%) was observed in the late–partial EMI schools. This is similar to Lo and Macaro's (2012) study in which they found that L2 use comprised over 96.8% of interactions in early–full EMI schools but that L1 use in three late–partial (MOI-switching schools) was higher, comprising over 97.6% of discussion.

The descriptive statistics show that teachers and students in early–full EMI schools tended to have more inter-sentential code-switching (in between sentences). Those in late–partial EMI schools tended to have more intra-sentential code-switching (within a sentence) although the difference here is not large when compared with the early–full EMI schools. This perhaps suggests that teachers and students in early–full EMI schools were experienced in EMI and thus could produce more completed utterances in English and inter-sentential code-switching. Those in the late–partial EMI schools were at the early transition of MOI change, and thus students could perhaps only produce phrasal utterances in English and intra-sentential code-switching with Cantonese as the matrix language.

In the macro level, although the observed late–partial EMI schools appeared to be slightly more teacher-centred, in which teachers contributed to over 68% of the interaction time and students contributed to less than 5% of the interaction time, the late–partial EMI schools seemed to be more interactive, given that teachers and students took more turns. This finding is similar to previous research which found that MOI-switching schools (equivalent to late–partial EMI schools in this study) were more interactive when the MOI changed from English to Cantonese (Lo & Macaro, 2012; Yip et al., 2007).

Given that the nature of science teaching is often monologic and that most science classrooms are teacher-centred in Hong Kong (Lin, 2006; Yip, 2003), it is unsurprising that in both types of schools, nearly 70% of all interactions consisted of teacher talk, with only a small percentage of student talk. However, there was more student talk in the early–full EMI schools than in the late–partial EMI schools. This is perhaps because students had more experience with EMI and better English proficiency, which allowed them to be more confident in responding to teacher questions.

Although the late–partial EMI schools seemed to be more interactive than early–full EMI schools, the average length of teacher turns (42.03 s < 49.29 s, respectively) and student turns (3.16 s < 8.48 s, respectively) were slightly shorter. The maximum length of student turns in the late–partial EMI schools was four times shorter (302.6 s < 1,192.1 s, respectively) than teacher turns in the early–full EMI schools. This could be explained by the fact that students encountered difficulties in expressing themselves using L2 (e.g., they only used English for scientific terms), leading to shorter answers and shorter turns when the MOI changes to L2, as Lo and Macaro (2012) indicated that there was a significant negative relationship between L2 use and student talk. It was found in their MOI-switching (late–partial) schools that students had shorter turns even though they had been studying through EMI for a few years.

In addition, findings from this study have revealed that teachers in both types of schools used a similar set of pedagogical functions during interaction (e.g., teacher informs, teacher self-utterances, teacher directs, and teacher replies) and offered similar types of feedback. In other words, the typical sequence was: 'affirmation-direct instruction' > 'extension by responsive questioning: focusing and zooming' > 'explicit correction-direct instruction' > 'constructive challenge'.

The results for pedagogical functions during interaction are similar to Lo and Macaro's (2012) findings, where 'inform' and 'direct' were the predominate teacher acts in most content-subject classrooms in Hong Kong. In the three observed MOI-switching schools, they found that the most frequent act was 'inform' (46.8–50.6%), and the second most frequent was 'direct,' which led to their observations of teachers' pedagogical skills where they found that teachers rarely adopted the kind of pedagogical scaffolding that would enable students to arrive at a deeper understanding of the concepts and processes required in academic subjects. Instead, the teachers were generally 'informing' students of the facts, as seen in the cases presented in this study from the early–full and late–partial EMI schools.

Previous studies have pointed out that science classrooms in Hong Kong are teacher-centred and that the role of the teacher seems to be transmitting knowledge to students (Lin, 2006; Lo & Macaro, 2012; Yip, 2003; Yip et al., 2007). To transmit knowledge, these teachers use monologic-style teaching techniques, such as informing, self-utterance, and directing. For instance, teachers inform students about a science phenomenon, provide definitions and explanations in a number of self-utterances (i.e., extended teacher informs) and direct students to their intended view of the science phenomenon. Teachers only reply to students who share the teacher's interpretation of the science phenomenon or whose responses fit the teacher's goals for the lesson. These features of transmitting the teacher's own view of the scientific story to students may reflect that fact that science teaching at senior secondary levels in Hong Kong tends to be teacher-centred, monologic, authoritative, and non-interactive. The authoritative, teacher-centred interaction patterns revealed in the data perhaps explain why the teaching styles of the EMI science teachers in the early–full and late–partial EMI schools did not vary dramatically. Thus, differences in this kind of 'teacher-centred, transmitting' teaching style cannot solely explain the observed differences between the schools.

The findings of this study are therefore different from those of previous studies conducted by Lo and Macaro (2012, 2015). Those studies claimed that the overall interaction patterns of schools with an earlier start and those schools with a later start of EMI were different in terms of their nature of classroom interaction. For example, Lo and Macaro found more and longer interactive sequences in classes where the MOI was changed from English to Cantonese. However, their findings were based on the situation before the launch of the fine-tuning language policy in 2010, at a time when there was a clear division between CMI and EMI schools. After the implementation of the fine-tuning policy under which all schools adopted EMI, differing only in whether they began using EMI earlier or later, this study found that the degree of similarity in the interactions between early–full and late–partial EMI schools was limited to the macro level, while there were substantial differences at the micro level.

When the classroom interactions were examined at the micro level in terms of the average length of teacher and student talk, the ratio between L1 and L2 use and the patterns of code-switching (inter- and intra-sentential) suggests a number of subtle differences between early and late EMI schools. In the early–full EMI schools, the average turn length of teacher talk and student talk was longer than that in the late–partial schools. This suggests that teachers and students use relatively more L2 than L1 in interactions, with the majority of interactions spoken in English. Teachers and students tend to code-switch between English and Cantonese at the sentence level (inter-sentential). In other words, in the early–full EMI schools, teachers tend to use English as the matrix language and embed within it Chinese phrases or words for direct translation. Teachers tend to use more lower-order questions, as compared with those in late–partial EMI schools, to recall students' previous learned science concepts. This is reflected in a shorter wait time between teachers' initiations and students' responses. This echoes the findings of Lo and Macaro's (2012) study, in which they found that the questions asked most frequently were factual questions, and when more L2 used, the more teachers talked in lessons, spoke longer at each turn, gave longer wait time for students to formulate answers to questions, and asked questions less frequently.

Unlike Lo and Macaro's (2012) observation in the MOI-switching schools, generally in the late–partial EMI schools, there were more (but shorter) student initiations and responses as well as more higher-order questions from the teachers to encourage students' cognitive thinking about science concepts. These questions were challenging enough to encourage students to resolve inconsistent views on science topics and develop scientific concepts themselves. Teachers provided less direct feedback to students than in the early–full EMI schools. Teachers and students used relatively more L1 than L2 in the interactions, with the majority of interactions spoken in Cantonese. This result echoes Lo's (2015) finding that content-subject teachers in Hong Kong used more L1 but in a skilful way.

One question that then arises: Why did the early–full EMI teachers use more lower-order questions? It could be that the students' English proficiency was limited such that they could not respond to higher-order questions. Teachers in the early–full EMI schools thus had to use lower-order questions to ensure that their students with poor English abilities were still able to interact in an English environment. This raises the question about the extent to which EMI is effective for content teaching in the late–partial EMI immersion context. Another possible reason could be that teachers believed that by using more lower-order questions, they were more likely to ensure that students mastered the teachers' authoritative scientific view. As explained in Chapter 4, the early–full EMI classrooms were slightly more teacher-centred, authoritative, and non-interactive than the late–partial EMI classrooms.

A further possible explanation for the reason why teachers in early–full EMI schools used lower-order questions may be because they expected students to answer in English. On the one hand, they may have believed that lower-order questions were more appropriate in a context where all interactions had to be conducted in English despite the limited English proficiency of their students. On the other hand, the late–partial EMI teachers possibly used more higher-order questions because they allowed their students to reply in Cantonese. Although the students' English proficiency was limited, teachers and students could interact through Cantonese, and thus the teachers perhaps felt that it was possible to use more higher-order questions.

The late–partial EMI teachers and their students tended to code-switch between English and Cantonese at the intra-sentential level, suggesting that teachers and students tend to insert English technical terms into Cantonese grammar and discourse. However, in both early–full EMI and late–partial EMI schools, code-switching between sentences is more frequent in Grade 11, suggesting that the extra year of EMI may help students to move from inserting single English noun phrases into their Cantonese to using more complete English sentences while interacting.

When the pedagogical functions of both teachers' and students' interactions were examined, it was found that the functions of interactive moves and the choices of teachers' questions were perhaps the two key indicators marking the differences between early–full and late–partial EMI schools.

The four most frequent pedagogical functions in the interaction sequences were inform, self-utterance (or extended teacher informs), direct, and reply. In addition to the fact that science teaching in Hong Kong is teacher-centred and authoritative, this may be the reason why, as mentioned earlier, in the early–full EMI schools, the teachers seem to help students less to develop higher-order thinking. Lo and Macaro (2012) found that teachers in both early–full and late–partial EMI (or MOI-switching schools) rarely adopted the pedagogical scaffolding that enables students to achieve a deep understanding of content concepts. This was the case in my study, where both early–full and late–partial EMI teachers rarely interacted with students skilfully in developing their understanding in science and language. The teachers in my study generally informed their students of factual information and tended to ask lower-order questions, namely recalling questions, to help students achieve basic understanding of concepts.

However, in my study, the late–partial EMI teachers seem to ask more cognitively challenging questions to stimulate students' higher-order thinking. Perhaps this is because the late–partial EMI schools are in a transitional phase of adopting EMI. These teachers used pure Cantonese as the medium of instruction or English technical terms inserted into Cantonese as the matrix language. In this way, teachers and students could work on cognitively challenging questions in Cantonese, using their rich semantic resources in their mother tongue to help promote higher-order thinking. This echoes Lo and Macaro's (2012) observation of the teachers' questioning techniques in the MOI-switching (late–partial) schools. They found that teachers and students freely switched back to Chinese when students had difficulties in answering questions.

These teachers are demonstrating the kind of bilingual pedagogical scaffolding that enables students to arrive at a deep understanding of science concepts (Lin, 2006). The use of Cantonese is likely to reduce students' anxiety greatly and lower the language barrier. Interactions in late–partial EMI schools are thus not surprisingly more interactive than those in early–full EMI schools. This echoes the findings from previous studies which have shown that MOI-switching schools are interactive and that the L1 can serve as a rich semantic resource to promote higher-order thinking (Lin, 2006; Lo & Macaro, 2012; Yip et al., 2007).

The subtle differences reviewed above reflect the fact that the interactions in early–full and late–partial EMI schools are different on a micro level. Perhaps this relates to the choice and extent of L1 used in the classroom, reflecting the fact that late–partial EMI science teachers in Hong Kong make judicious use of Cantonese to facilitate their students' learning, taking into account their English abilities (Lin, 2006; Lo, 2015). This finding echoes Lo and Macaro's study (2012), which found that lessons with more Chinese as the MOI tended to be more interactive than those with more use of English.

9.1.3 Research Question 3: What are the differences and similarities in classroom interaction between Grades 10 and 11 in each group?

During the first year of the senior secondary science classroom (Year 11), the students at the late–partial schools in my study had switched from Cantonese to English medium, and from studying one integrated science subject into different science subjects (physics, chemistry, biology). Comparing the interaction in Grades 10 and 11, regardless of school type, showed that both types of schools shared a number of similarities in terms of interaction time, length of student turns, and ratio between L1 and L2 use.

In both types of schools, there was a small decrease in the percentage of interaction time, a decrease in the maximum length of student turns and a small increase of L1 use from Grades 10 to 11. Similar patterns were observed in L1 and L2 use. Teachers and students used relatively less L2 in interactions, with a similar ratio between L1 and L2 used in both teacher and student talk. In the late–partial schools, teachers and students contributed 44% L2 and 56% L1 at Grade 11, and 48% L2 and 53% L1 at Grade 10. These findings are unlike those from Lo and Macaro's study (2012), in which they found that their MOI-switching schools had a dramatic increase (20–70%) in L2 use a year after the MOI change. In this study, I did not observe any such obvious change in the MOI from L1 to L2 in my two types of EMI schools.

Teachers and students tended to insert English technical terms into Cantonese interactions. There was more L1 use (46%–>54%) at Grade 10 than Grade 11 in late–partial EMI schools, which suggests that students and teachers in Grade 11 still use an extensive amount of L1 in their teaching and learning after the first year of EMI experience. More code-switching between sentences (inter-sentential) was observed in Grade 11 in both types of schools. This is perhaps because students and teachers with more experience in EMI have more confidence producing English utterances.

Lo and Macaro (2012) showed that there were noticeable changes in teacher–student interaction patterns when the MOI changed substantially from the students' L1 in Grade 9 to L2 in Grade 10. But from the data collected in this study, I did not see such changes in the four late–partial EMI schools from Grade 10 to 11. The teacher–student interaction patterns in both early–full EMI and late–partial EMI (or MOI-switching) schools were similar in this study. One possible reason is that two of the four late–partial EMI schools used more Cantonese as their instructional language together with English despite their stated policy as EMI schools.

Although the language policy and the school website indicated that English was the official instructional language, teachers and students in both the early–full and late–partial EMI schools used English and Cantonese. As regards the language of the written materials, I observed that the textbooks and notes used in the four early–full and four late–partial EMI schools were entirely in English, but two of the late–partial schools provided written materials partially in Chinese, including Chinese glosses of difficult vocabulary items.

From the post-observation interviews with the teachers, I established that the amount of Cantonese used in the classrooms largely depended on the students' English proficiency. Different schools might change the proportionality of their medium of instruction (i.e., using a different amount of L1 in the classroom). In early–full EMI schools, a number of lessons were observed in which the majority of the teachers and students interacted in English with some degree of Cantonese.

In late–partial EMI schools, the majority of the teachers and students inter-acted mostly in Cantonese with English technical terms being used. In school G, I observed a class in which students assigned to either L1 or L2 instruction were studying in the same class due to a shortage of teachers which prevented the school from offering separate English medium and Chinese medium classes for each science subject.

The reasons for an increase in L1 use from Grades 10 to 11 may have to do with examinations. In the post-observation interviews, teachers explained that students in Grade 11 were planning to switch back to the L1 medium for their public examinations in Grade 12. Thus, the teachers used more L1 in the class to accommodate these students' needs. This result echoes Lin's (2006) and Lo's (2015) findings that teachers adopted bilingual pedagogical strategies and used L1 to make science content more accessible to students with poor English abilities. Other research has disclosed the gap between policy and implementation through actual practice in the recent fine-tuning of the Hong Kong language policy (Chan, 2014; Evans, 2009; Poon, 2013). In other words, it seems that there was a subtle change in the MOI in oral and written materials in the late–partial EMI schools. However, this change in the MOI from English-only to mixed-mode instruction did not result in different classroom interaction patterns.

My data also revealed subtle differences between the two types of schools between Grades 10 and 11. In the early–full EMI schools, the difference in the mean length of teacher turn was statistically significant. There was more teacher talk but less student talk in Grade 11 than in Grade 10. In the late–partial EMI schools, however, teacher and student talk was shorter in Grade 11 than in Grade 10. The average length of teacher turns in the early–full EMI schools was much shorter than the late–partial schools.

By comparing Grades 10 and 11, the average length of teacher turns was longer in Grade 11 (38.67 s compared with 47.24 s, respectively) but the student turns were shorter (3.31 s compared with 3.00 s, respectively) in the late–partial EMI schools. In addition, the maximum length of student turns in Grade 11 was shorter in both types of schools. The late–partial schools in this study contradicted Lo and Macaro's (2012) observation in MOI-switching schools, where they found that student turns were longer in terms of time taken to utter their thoughts.

Given that the science curriculum is more difficult in Grade 11 than in Grade 10, and that students have become used to EMI teaching, teachers in early–full EMI schools might use many shorter turns to elicit science explanations. I had expected to find that students in early–full EMI schools would interact more in Grade 11, but the data show that student talk in Grade 11 was much less than in Grade 10 (4% compared with 17%, respectively). However, there was no major change in the late–partial EMI schools. In this study, teachers and students interacted mainly in Cantonese. This result is different from that of Lo and Macaro's (2012) study, in which they concluded that the change of MOI towards English could cause lessons to become more teacher-centred, with fewer opportunities for negotiating meaning and scaffolding. In this study, even though late–partial EMI students in Grade 11 had changed their MOI to English, they were able to interact freely in a comfortable language environment while learning more abstract and difficult science topics.

In addition, with pedagogical moves, there were more 'checks'[1] than 'self-utterances'[2] in Grade 11 in the late–partial EMI schools, but this pattern was reversed in Grade 10 in the early–full EMI schools. Perhaps the teachers in Grade 11 in the early–full EMI schools made more effort to ensure that students shared the same understanding of the science topics. In the late–partial EMI schools, in Grade 11, there was more 'self-utterance' than 'direct,'[3] but this pattern was reversed in the early–full EMI schools. Perhaps the early–full EMI teachers made more effort to provide explanations to the students when the topic was abstract and difficult.

9.1.4 Research Question 4: What are teachers' perceptions of teaching science through EMI?

From the interviews with the early–full and late–partial EMI school teachers, 11 re-occurring issues were identified: (1) overall views on teaching experience in EMI science classrooms; (2) language challenges in teaching science through English; (3) specific language challenges in biology/chemistry/physics; (4) coping strategies; (5) dealing with students with difficulty in communicating in English; (6) when to use L1; (7) questioning techniques in EMI classrooms; (8) vocabulary types in EMI science classrooms; (9) teacher's expectations of students' English level; (10) teacher's expectations of students' science learning; and (11) other teaching challenges related to the new science curriculum.

The teachers from both the early–full and late–partial EMI schools shared many similar views about EMI teaching, such as their preferred instructional language and identifying the most important and demanding English language skills for learning science. This finding of common attitudes was, however, unexpected. Teachers generally welcomed a mixed-mode of instruction to allow for the optimal use of L1 in explaining cause-and-effect relationships, giving manageable explanations to students with limited English, and consolidating students' knowledge by making connections with everyday examples. However, the teachers' views on the reasons for adopting EMI differed. Teachers from the late–partial EMI schools strongly believed the MOI was just a tool for transmitting knowledge rather than a way of preparing students to learn a second language, whereas some biology teachers from early–full EMI schools believed that they needed to help students with their English in science. When asked when teachers should use L1, the majority of teachers said that they would use L1 for students with limited English abilities. This echoes the findings of previous studies (Lin, 2006; Lo, 2015; Lo & Macaro, 2012; 2015; Tavares, 2015) in Hong Kong and in South Africa (Probyn, 2006, 2009), where content-subject teachers in L2 classrooms judiciously used L1 to facilitate their students' learning, taking into account their English language ability.

Teachers in late–partial EMI schools generally used a range of question types, such as recalling, higher-order, and reasoning questions, to promote student's science learning, but, overall, teachers tended to use more low-order questions. Yip (2003) also found that junior science teachers in Hong Kong were generally skilful in questioning but focused on questions with a lower cognitive demand. This was due to the time constraints on teaching, the pressure to cover the examination syllabus within a tight schedule (Yip, 1999), the low English proficiency of students, and the poor communication skills of some EMI science teachers (Yip, 2007a, 2007b). However, in my study, I found that biology teachers paid more attention to students' language problems. Biology teachers from early–full and late–partial

EMI schools shared a number of coping strategies to address their language issues and improve their students' English abilities. Although these biology teachers felt they should help with students' English, none of them (from the interviews) said they had the opportunity and responsibility.

Both early–full and late–partial EMI teachers said that the major challenges of teaching science through EMI were students' analytical, reading, and communication skills. They recognised that students had problems understanding examination questions and presenting science ideas in writing or speaking. They cannot present coherent science explanations in English using adequate vocabulary and expressions. When learning science, students encountered three types of vocabulary: basic English vocabulary, academic vocabulary, and scientific or technical vocabulary. In the teacher interviews, teachers said students had more problems with basic English vocabulary because they misinterpreted its meanings in science and confused word meaning with other meanings they had learned in English lessons. Students also mixed up technical vocabulary that had different meanings in different scientific contexts. Teachers believed poor vocabulary skills affected students' writing abilities and communication skills, especially when students were asked to write an essay explaining a scientific phenomenon.

When asked about their expectations of students' English levels, the majority of science teachers said they did not have any expectations. These science teachers believed that language is just a tool to facilitate learning. Perhaps there is a gap here in science teachers' perceptions about EMI and their actual responsibilities in teaching students various science concepts (Probyn, 2006, Yip, 2003). In addition, research in teacher education programmes has shown that teacher training does not prepare science teachers to develop the necessary pedagogical skills for integrating both content and language (Othman & Mohd. Saat, 2009), nor does the training help them develop their identity as a teacher of both academic content and academic language (Coyle et al., 2010). Wannagat (2007) compared the classroom interactions of history lessons between EMI and CLIL (Content Language Integrated Learning) settings. From his observations, the role of language in learning goals is not clearly explicated in EMI, and the language issue is largely ignored in curriculum development and teacher training. Thus, he believed EMI is less successful than CLIL because there is no language focus in the EMI curriculum, although there is more English exposure for students in EMI.

When asked about their expectations of students' science achievements in EMI classrooms, the majority of science teachers said they hoped that their students would master key scientific concepts, develop strong observational skills, improve their analytical and critical thinking skills, and ultimately show a strong interest in science and its applications in our society.

On two points, there were two statistically significant differences in the questionnaire responses from teachers in the early–full and late–partial EMI schools: (1) the most important English language skills for students to study science in English medium, and (2) teachers' language awareness in teaching science through English. The late–partial EMI teachers used more Cantonese in science-teaching activities. Compared with early–full EMI teachers, these late–partial

EMI teachers believed that the English language skills that students needed in order to learn science were more demanding but that these language skills were less important. Scientific terms, especially in English, are usually longer and more complex compared with normal daily-life English vocabulary. Hence, students may find it hard to catch up when they are asked to pronounce or memorise these words. For the early–full EMI, however, teachers thought the reverse. In addition, the late–partial EMI teachers had lower language awareness than those in early–full EMI schools, suggesting that they paid less attention to their students' language issues. However, the late–partial EMI teachers reported that they had more language challenges, suggesting that they believed their students encountered a number of language challenges. However, the late–partial EMI teachers reported that they used coping strategies less frequently and found these strategies less useful than teachers in the early–full EMI schools. The late–partial teachers also had a higher positive reflection towards their teaching practices in the EMI context, suggesting that EMI has impacted their teaching.

A number of teacher demographic characteristics are correlated (e.g., the school's current MOI; teachers' self-reported English overall scores; scores in listening, reading, writing and speaking; and EMI training). Teachers with lower self-reported English scores, particularly those with poor speaking skills, tended to use Cantonese more frequently in science-teaching activities. The late–partial EMI teachers believed that English language skills were more important than the early–full teachers. The early–full EMI teachers with higher self-reported English scores had higher language awareness than the late–partial EMI teachers. Teachers with self-reported EMI training, fewer teaching experiences at senior levels, or those who believed English language skills were important had a positive teaching belief towards EMI. Those teachers who reported that they used coping strategies frequently and also found these strategies more useful than their peers.

9.1.5 Research Question 5: What are students' perceptions of learning science through EMI?

The interview analysis identified seven recurrent themes from the focus group student interviews: (a) language choice for interactions; b) language challenges for vocabulary in science subjects; (c) coping strategies related to vocabulary; (d) language challenges related to writing and reading in science; (e) coping strategies for writing and reading in science; (f) strategies related to external help; and (g) positive and negative attitudes towards EMI.

Students from both the early–full and late–partial schools shared many similar views about their EMI experiences. Students in both groups welcomed EMI instruction. However, while students in the late–partial EMI schools saw EMI as an opportunity to improve their English, those in the early–full EMI schools believed that EMI demotivated their learning of science. In the interviews, both groups of students highly valued the role of L1 in learning science through English. They identified the challenges of learning vocabulary in science subjects, noting that there were too many specific science terms to remember, as well as words with multiple meanings, connotations, and associated meanings. Some students found it difficult to use scientific words to convey simple meanings due to problems conjugating parts of speech or forming different word forms. Students used a number of coping strategies, including rote memorisation, using synonyms, and understanding associated meanings in similar science contexts.

Students in both groups also pointed out a number of language challenges related to writing and reading in science. Students found it very difficult to communicate and explain their scientific ideas in a complete sentence in English. In my study, there was clear evidence of a discrepancy between what students understood in Cantonese and what students produced in English. Students also reported that they struggled to identify relevant key scientific ideas, guess the meaning of an unfamiliar word (e.g., a disease name), and understand the nature of examination question words. They stressed the need for explicit teaching about the kind of sentence patterns and structures expected in their science writing. The students interviewed said perhaps it was the responsibility of their science teacher to focus on key concepts rather than to spend a lot of time on elaborations or explaining everyday examples. Students also sought help from tutorial centres.

There was a statistically significant difference in students' survey responses in early–full and late–partial schools for the following four aspects: (1) their language choices in science-teaching activities, (2) the most important English skills, (3) their language challenges, and (4) the frequency of coping strategies. Between Grades 10 and 11, there was a significant difference in the early–full school group in two aspects: students preferred more English as the language choice in science-teaching activities and they found coping strategies useful in adapting to language challenges.

Students from the late–partial EMI schools had a higher mean score in the four aspects mentioned above, suggesting that they used more Cantonese in science classrooms. Students in the late–partial EMI schools also reported lower self-concepts in science, suggesting that fewer of the students in the late–partial schools were confident in their ability to do well in science. The late–partial EMI students also rated the same set of language challenges higher than the early–full EMI students, yet they used coping strategies less frequently and found them less useful than students in the early–full EMI schools. The late–partial EMI students found English to be demanding in areas where they needed to use vocabulary to express abstract concepts in science.

Students who used more Cantonese in science-teaching activities believed that English language skills for learning science were demanding. They reported that they faced more language challenges while learning science in English. Those who rated their demanding English skills of higher importance also reported more language challenges and more frequent use of coping strategies.

These results could explain the findings of Yip et al. (2007), in which 6,000 EMI junior secondary science students reported a lower self-concept in science (e.g., students' view of their ability to do well in science; Wilkins, 2004) than those taught in CMI. Although EMI students had a lower self-concept and self-competence in science, they performed better in science achievement tests than many of the CMI students. Students who think the English skills for learning science are important had a stronger positive belief in EMI. These students also found coping strategies more useful. They preferred to use more English in science-teaching activities. Students who had more positive reflections on EMI also believed that they would do better in science subjects.

9.2   Contributions to theory and practice

9.2.1   *Contributions to theory*

The present study contributes to current theories in a number of ways. First, it extends exchange structure analysis (IRF sequences) to a macro level. IRF analysis demonstrates the role of the teacher's modified input in talk and students' thematic understanding in science. The sheer increase in the number of IRF extended exchanges, teacher's modified input in scaffolding, more students' responses, and teacher's feedback along with the increased opportunities to negotiate meaning and acquire language in science are highly facilitative for EMI students to acquire both content and language and to attain a better learning belief and self-concept in science.

Second, this study connects Mortimer and Scott's (2003) communicative framework to the micro-level understanding of how teachers interact with students in different types of EMI science classrooms. The examples shown in Chapter 5 from the four biology classrooms reflected Mortimer and Scott's (2003) insight that learning in science classrooms consists of co-constructing meaning through language. Such learning may take place through collaborative dialogue in the form of extended IRF exchange between teachers and students or monologically by a teacher through modified talk. In the analysis of classroom interactions in Chapter 5, I summarised the features of IRF interactions at the micro level to provide a comprehensive account of how the teacher interacted with the students to demonstrate the ways science teachers in EMI classrooms make scientific ideas more accessible to students and provide students with opportunities to learn the content and language. The IRF features observed (and summarised in Table 5.8) in the four biology EMI science classrooms can provide a supplement to Mortimer and Scott's four-part communicative framework.

Third, this study applies SLA theory (i.e., interaction hypothesis) to content-based instruction contexts. This provides empirical evidence to support Long's (1983) interaction hypothesis in SLA theory, which argues that students need to negotiate meaning through meaningful interaction with a teacher in order to acquire the form as well as the content knowledge.

Fourth, this study shows that teachers' modified input plays an important role in teacher–student interaction. The decision as to whether a teacher selected a student's answer to expand for further discussion or to acknowledge without follow-up was controlled by the teacher. Such decisions are a determining factor in shaping the nature of what Mortimer and Scott (2003) recommend as communicative approaches in science classrooms. The evidence presented in the two analyses in Chapter 5 show how teacher interactions with students help develop students' thematic understanding of science. Such analysis would provide insights into how the macro level (different communicative approaches in science classrooms) and micro level (IRF patterns) connect.

9.2.2   *Contributions to practice*

The present study hopes to contribute to current practice in a number of ways. First, it sheds light on MOI issues in the sense that it details two aspects of EMI experience in different types of EMI schools. That is, it details both classroom interaction and teacher and student attitudes about their teaching and learning experiences in both early–full and late–partial EMI schools.

Secondly, the study suggests that the adoption of EMI with the aim of providing students with more English learning environments to enhance their exposure to L2 may be justified by a school-based MOI policy, as seen in the cases of late–partial EMI schools where teachers and students chose to have mixed-mode instruction and adopted L1 use for classroom interaction. For example, I illustrated this with three examples of biology teachers asking higher-order questions, explaining words with similar meanings, and using synonyms to describe a biological concept in Chapter 5. This is particularly important in Hong Kong where the overwhelming majority of the population is Chinese, and Cantonese is often the only language spoken at home and used in the community.

Third, teachers' and students' perceptions on their teaching and learning experience in EMI science classrooms may reflect the numerous educational and practical issues that have been identified in EMI or L2 instruction in Hong Kong and worldwide. Some of these issues may include

- The unclear role and pedagogical functions of optimal L1 use in science teaching
- Different views towards the teaching responsibility of teaching both content and language (i.e., many science teachers believed EMI is just a tool for transmitting knowledge rather than preparation for students to learn a second language)
- Teachers' poor awareness of students' language problems
- Students' inadequate levels of English for learning science through English
- Students' difficulties in expressing science ideas through English
- Students' poor vocabulary knowledge (basic English, academic, scientific)
- Students' diversity (different science backgrounds, English levels)
- Students' orientation to rote-learning
- Students' low motivation for learning science.
- Students' experiences of being overloaded with topics in the new science curriculum
- Unsystematic teaching of curriculum topics
- The expectation that teachers teach outside of their speciality (e.g., teaching basic maths or other related science concepts).

Given these many issues, perhaps there is a need to develop adequate teacher training to equip EMI teachers with practical solutions. Such professional development would be needed to equip these EMI teachers with the necessary pedagogical skills to integrate content and language (Othman & Mohd. Saat, 2009), and develop their identity as CLIL teachers by combining content and language learning in their planned curriculum (Coyle et al., 2010).

9.3   Summary

This chapter has summarised the key findings of this study and brought them together to provide an overall picture of secondary science teaching and learning in the two types of EMI classrooms, shedding light on how we can most effectively inform teaching and learning processes in EMI classrooms. Drawing on the key findings, it has answered all the research questions stated in the beginning of the book and summed up the study's contributions in terms of both theory and practice. In the next chapter, I will conclude the book with a discussion of the implications and applications for the professional growth of science teachers as well as other content-subject teachers, and offer teaching recommendations for science in various EMI situations.

Notes

1 'Check' is realised by a closed class of questions such as 'O.K.?', 'Finished?' and 'Ready?' It helps the teacher to know the progress of the lesson and usually occurs when the teacher wants to move from one task or topic to another.
2 'Self-utterance' is realised by statements or rhetoric questions. It refers to those utterances that the teacher speaks to himself/herself. It is an extended teachers' 'informs'. It does not aim at eliciting students' responses.
3 'Direct' elicits non-verbal responses. It can be realized by questions or imperatives. For example 'Sit down' or 'Turn to page xx'.

References

Chan, J. Y. H. (2014). Fine-tuning language policy in Hong Kong education: Stakeholders' perceptions, practices and challenges. *Language and Education*, *28*(5), 459–476.

Coyle, D., Hood, P., & Marsh, D. (2010). *CLIL: Content and language integrated learning*. Cambridge: Cambridge University Press.

Evans, S. (2009). The medium of instruction in Hong Kong revisited: Policy and practice in the reformed Chinese and English streams. *Research Papers in Education*, *24*(3), 287–309.

Lin, A. M., (2006). Beyond linguistic purism in language-in-education policy and practice: Exploring bilingual pedagogies in a Hong Kong science classroom. *Language and Education*, *20*(4), 287–305. http://doi.org/10.2167/le643.0

Lin, A. M., & Wu, Y. (2015). 'May I speak Cantonese?'–Co-constructing a scientific proof in an EFL junior secondary science classroom. *International Journal of Bilingual Education and Bilingualism*, *18*(3), 289–305.

Lin, L. H. F., & Morrison, B. (2010). The impact of the medium of instruction in Hong Kong secondary schools on tertiary students' vocabulary. *Journal of English for Academic Purposes, 9*(4), 255–266. http://doi.org/10.1016/j.jeap.2010.09.002

Lo, Y. Y. (2015). A glimpse into the effectiveness of L2-content cross-curricular collaboration in content-based instruction programmes. *International Journal of Bilingual Education and Bilingualism, 18*(4), 443–462.

Lo, Y. Y., & Macaro, E. (2012). The medium of instruction and classroom interaction: Evidence from Hong Kong secondary schools. *International Journal of Bilingual Education and Bilingualism, 15*(1), 29–52.

Lo, Y. Y., & Macaro, E. (2015). Getting used to content and language integrated learning: What can classroom interaction reveal? *The Language Learning Journal, 43*(3), 239–255.

Long, M. H. (1983). Native speaker/non-native speaker conversation and the negotiation of comprehensible input. *Applied Linguistics, 4*(2), 126–141.

Marsh, H.W., Hau, K.T., & Kong, C. K. (2000). Late immersion and language of instruction in Hong Kong high schools-achievement growth in language and nonlanguage subjects. *Harvard Educational Review, 70*(3), 302–346.

Mortimer, E., & Scott, P. (2003). *Meaning making in secondary science classrooms.* England: Open University Press.

Othman, J., & Mohd. Saat, R. (2009). Challenges of using English as a medium of instruction: Pre-service science teachers' perspective. *The Asia-Pacific Education Researcher, 18*(2), 307–316.

Poon, A.Y.K. (2013). Will the new fine-tuning medium-of-instruction policy alleviate the threats of dominance of English-medium instruction in Hong Kong?. *Current Issues in Language Planning, 14*(1), 34–51.

Probyn, M. (2006). Language and Learning Science in South Africa. *Language and Education, 20*(5), 391–414. http://doi.org/10.2167/le554.0

Probyn, M. (2009). 'Smuggling the vernacular into the classroom': Conflicts and tensions in classroom codeswitching in township/rural schools in South Africa. *International Journal of Bilingual Education and Bilingualism, 12*(2), 123–136.

Swain, M. (1995). Three functions of output in second language learning. *Principle and practice in applied linguistics: Studies in honour of HG Widdowson, 2*(3), 125–144.

Tavares, N. J. (2015). How strategic use of L1 in an L2-medium mathematics classroom facilitates L2 interaction and comprehension. *International Journal of Bilingual Education and Bilingualism, 18*(3), 319–335.

Wannagat, U. (2007). Learning through L2 – Content and Language Integrated Learning (CLIL) and English as Medium of Instruction (EMI). *International Journal of Bilingual Education and Bilingualism, 10*(5), 663–682.

Wilkins, J. M. (2004). Mathematics and science self-concept: An international investigation. *The Journal of Experimental Education, 72*(4), 331–346.

Yip, D. Y. (1999). Implications of students' questions for science teaching. *School Science Review, 81*(294), 49–53.

Yip, D. Y. (2003). *The effects of the medium of instruction on science learning of Hong Kong secondary students.* Doctoral dissertation: University of Nottingham.

10 Conclusion

10.1 The nature of classroom interactions in EMI classrooms

As explained in Chapter 8, the data collected for this project reveals that classroom interactions in EMI science classrooms were authoritative and less interactive, very often with teachers dominating the interactions. Interactions between teachers and students were limited, with simple initiation-response-feedback (IRF) patterns, which reduced the opportunities for teacher–student negotiation of meaning and understanding. Students seldom initiated interactions and were mainly invited by teachers to give a response to show their understanding of the science topics. This suggests that the nature of classroom interaction in EMI science classrooms tends to become more teacher-centred with students' participation being inhibited when more English is used as medium of instruction (MOI), as it was the case in early–full schools involved in this study. Students' language proficiency and teachers' pedagogical skills may be two major reasons which could account for the limited teacher–student interaction. Students with limited oral English proficiency might have limited ability to negotiate meaning with their teachers, and therefore their teachers might have to use lower-order questions to ensure that their students with poor English ability could still interact in an English environment.

Teachers trying to teach science through English failed to adopt appropriate pedagogical skills for delivering content through English (e.g., higher-order questions, clarification requests, confirmation checks and signals, comprehension checks). These skills would enable students to arrive at a conceptual understanding in science, thereby promoting their deeper understanding of the science concepts and process. Instead, the teachers in this study tended to adopt lower-order pedagogical scaffolding, only 'informing' their students of the facts and subsequently encouraging students' rote-memorisation of factual information.

10.2 Comparing interactions in early–full and late–partial EMI *classrooms*

The early–full and late–partial EMI science classrooms showed similar interactional patterns at the macro-level (e.g., percentages of interaction time, distribution of time between teacher and student talk, and frequency of pedagogical functions).

DOI: 10.4324/9781003001454-11

However, when explored from the micro-level, it became apparent that the nature of the interactions was different. In late–partial EMI schools, there were generally more (but shorter) student initiations and responses and more use of higher-order questions from the teachers but less direct teacher feedback to students. Teachers and students used relatively more L1 than L2 in the interactions, with the majority of interactions spoken in Cantonese. Teachers and students freely switched back to Chinese when students had difficulties answering questions in the late–partial EMI schools. In this way, L1 could serve as a rich semantic resource to promote higher-order thinking. For example, teachers could ask higher-order questions, explain words with similar meanings, and use synonyms to describe a science concept in L1 (see examples in Chapter 5, Section 5.5.2).

Thus, these teachers seemed to ask more cognitively challenging questions through L1 to stimulate students' higher-order thinking. These teachers used Cantonese as the MOI with English technical terms. In this way, teachers and students worked on cognitively challenging questions in Cantonese, using their rich semantic resources in their mother tongue to help promote higher-order thinking. The use of Cantonese likely reduced students' anxiety and lowered the language barrier.

10.3   Comparing interactions in Grades 10 and 11 in each school group

The data showed less interaction time, shorter maximum length of student turns, and more L1 use in Grade 11 than in Grade 10 in both types of school. Perhaps this was because, with the advanced level and more abstract and difficult science concepts, teachers and students become less interactive in Grade 11. This was observed mainly in the early–full EMI schools where teachers and students had to interact mostly in English. However, students in late–partial EMI schools were able to interact freely in a comfortable language environment when learning more abstract and difficult science topics. This may explain why there was an observable increase in L1 use from Grades 10 to 11 as teachers used L1 to make science content more accessible to students with poor English abilities.

10.4   Teachers' perceptions of EMI teaching in science classrooms

The teachers in both the early–full and late–partial EMI schools shared similar views about their EMI teaching experiences. They welcomed a mixed mode of instruction to allow optimal use of L1 in explaining and providing feedback to students with limited English. However, the late–partial EMI teachers reported that they used more Cantonese when teaching science. Although the late–partial EMI teachers said the English language skills for learning science were demanding, they reported that these English language skills were less important for learning science. This result was also observed in the attitudes of students in the late–partial schools.

The late–partial EMI teachers had lower language awareness in teaching science through English. With regard to questioning techniques, the early–full EMI teachers preferred to use recalling questions, while the late–partial EMI teachers preferred to use reasoning questions (i.e., mediating through Cantonese). Although the late–partial teachers had more positive reflections about EMI science teaching, they reported that their students faced more language challenges. The late–partial students also reported they used coping strategies less frequently and found these strategies less useful.

10.5   Students' perceptions of EMI experiences in science classrooms

Students in both the early–full and late–partial EMI schools shared similar views about many aspects of their EMI experiences, both welcoming EMI but also identifying certain caveats. The late–partial EMI students used more Cantonese in science learning activities than the early–full EMI students. All students highly valued the role of L1 in learning science through English. They mentioned the challenges of learning vocabulary in science subjects, noting that there were too many specific science terms to remember, and that many words had multiple meanings in English, or different connotations and associated meanings outside of science. Students also found it very difficult to convey a complete meaning in science explanations, suggesting a discrepancy between what students understood in Cantonese and what they could produce in English.

In terms of English language skills, the late–partial EMI students believed the English language skills needed for learning science were demanding. However, many of the students thought these skills were not as important as did students in the early–full EMI schools. This indicates that perhaps students in late–partial EMI schools were not expected to have fully adopted EMI as early–full students did. The reason for this may be that in some late–partial EMI schools, students with limited English ability would opt to sit the Chinese version of the public examination at the end of their secondary education. This attitude towards English science learning skills was also observed among teachers in the late–partial EMI schools. The late–partial EMI students believed they were less capable in science and reported that they encountered more language challenges in the EMI classroom. Yet, the late–partial EMI students did not find coping strategies useful and thus used them infrequently. As a result, the late–partial EMI students saw EMI as an opportunity to improve their English, but the early–full EMI students believed that EMI demotivated their learning of science.

Among these results, one notable finding is that variation in the quality of classroom interaction as measured by a number of linguistic cues (e.g., IRF sequence, interaction time, wait time, number of turn-taking, distribution of talk, types of questions) seemed to be related more to the ability of the teacher to adjust his or her language to the needs of the student and to scaffold the students' exploration of the concepts introduced than to whether the students were of late or early immersion. This suggests that the most effective way of promoting deep learning in EMI science classrooms would be to provide training and support for science teachers in developing rich interaction (e.g., extended IRF sequence, more students' initiations, use of high order questions) in their classes.

10.6   Limitations of this study

10.6.1   *Sampling*

This study employed purposive non-probability sampling. Thus, the reported findings are not representative of the Hong Kong school population, nor can they be statistically generalised to that population. Moreover, it was not possible to control for the fact that participants might be receiving language support outside of their classrooms and that this could have had an effect on participants' performance in the classrooms observed. For example, many students in Hong Kong receive additional English support at tutorial centres where students can study both English and science in Cantonese, with study materials provided in English.

The number of lessons observed for this study and the number of participating schools was rather small. This limitation was due to the time and resources available for a doctoral study. In the research design phase, the eight participating schools and the lessons observed were carefully and purposely sampled so that they could be reasonably representative and be 'comparable' of the two types of EMI schools. Therefore, the two types of EMI schools studied included a range of students with similar academic ability and English proficiency. The study would have benefited from the inclusion of more lessons and more teachers. This would increase the generalisability to the population of most EMI secondary schools in Hong Kong. Nevertheless, the actual sample was by no means an insignificant one.

10.6.2   *Teacher and student effects*

The naturalistic observation design of this study has inevitably brought with it some limitations. For instance, it is difficult to select classes in Grades 10 and 11 from both early–full and late–partial EMI schools where students' English abilities and academic achievement are the same. I could only control the selection of schools according to their banding, which is based on students' attainment tests prior to admission. Thus, it was not possible to control for the effect of individual students and individual teachers in the classrooms observed as participation was voluntary. This might have affected the internal validity of the study as one could argue that the observed classroom interactions could also be the result of individual learning styles among students and different teaching styles among teachers. Individual factors such as underlying academic abilities and socioeconomic status would ideally need to be taken into account or controlled for. However, it is hoped that the inclusion of both early–full and late–partial EMI schools to provide a comparison in the design has alleviated such effects. A longitudinal study exploring the quality of classroom interaction with the same cohort of students across a number of years after the adoption of EMI would be a more valid and rigorous research design to use in addressing the research topic.

10.6.3 Teacher's and students' English proficiency levels

Students' and teachers' English proficiency could be a factor in explaining the differences across schools. Teachers self-reported their English abilities and students reported their school English achievement in the last academic year as a sub-indicator. Teachers' and students' English proficiency was not measured during their interactions. The measure of students' and teachers' English proficiency would provide clear evidence as to whether the observed classroom interactions were affected by the students' and teachers' English abilities, or whether there were other factors such as the teachers' teaching beliefs and the students' perceptions of EMI learning that played a role in shaping their classroom interaction.

10.6.4 Teachers' and students' L1 use

The 33 lessons observed in this study showed that even though the use of English was expected in these classrooms, the late–partial EMI teachers were aware of their L1 use and used English when they felt it was appropriate to their students' L2 proficiency and academic ability. The student interviews revealed insights from students about whether a teacher's use of L1 could help them grasp science content and then express their science ideas through English. Future research could examine how schools implement EMI courses from partial to full EMI instruction and when the L1 can most effectively be used in a science lesson for accessing content knowledge and facilitating L2 learning. For instance, I illustrated the three examples of biology teachers asking higher-order questions, explaining words with similar meanings, and using synonyms to describe a biological concept.

10.6.5 Non-verbal language in learning science

The goal of this research has been to examine the nature of classroom interactions in different types of EMI classrooms. The primary focus has thus been on how content and language are developed through interactions. However, other research has suggested that science teachers go beyond language (oral or written), drawing on multiple semiotic systems in the form of multimodal resources to explain abstract science concepts and facilitate the deconstruction of science knowledge (Lo, 2015; Lo & Macaro, 2012). Multimodal resources include symbols, graphs and diagrams, mind maps, tree diagrams, flow charts, crossword puzzles, whiteboards, and video games. Due to the scope of this study, I was not able to analyse in detail the relationship between language and multimodal resources in explaining scientific ideas. Nor was I able to examine the ways in which science teachers use multimodal resources to explain abstract scientific concepts during their interactions with students. In future research, it would be worth exploring the use of these semiotic resources to identify whether and how they help L2 learners unpack the technical, dense, abstract discourse patterns which characterise scientific discourse.

10.6.6 *Time and resources*

As a result of funding structures, the planning, execution, and reporting of this study were constrained to a period of three years, which made a longitudinal approach impossible. As a result, compromises were made, with data analysis restricted to cross-sectional, naturalistic observations of 33 lessons observed in different EMI classrooms in Hong Kong, pre-and post-observation interviews with teachers, and post-observation focus group interviews with students and questionnaires.

10.7 Contributions of the study

Having presented the findings from the study, this section summarises the contributions of the study for education in the Hong Kong context, research methodology, theory, and pedagogy. The section then offers suggestions for future research and ends with concluding remarks and recommendations.

10.7.1 *Contextual contributions*

1 This study adds additional evidence to previous research findings regarding the MOI policy in Hong Kong, where schools use either Chinese or English as their MOI. The results of previous studies may not reflect the current MOI situation in Hong Kong, since English is currently used in nearly all secondary schools to teach science subjects, whether the school is CMI or EMI. In addition to the fact that such policy had changed before this study began, there is a lack of research on the MOI situation in Hong Kong secondary schools following the most recent policy change.

2 This study offers recent experimental proof of the effects of the Hong Kong government's fine-tuning policy on the teaching and learning process in secondary schools. Classroom interactions are also observed in the science context, including physics, biology, and chemistry, to examine the MOI situation and its impact. According to Chan (2014) and Poon et al. (2013), the fine-tuning MOI policy brought limited educational value to the teaching and learning processes in secondary schools. For previously CMI students, especially those with poor English proficiency, it is doubtful whether they can learn as effectively in an EMI environment as they did in CMI classrooms.

3 This study suggests that the adoption of EMI might not ultimately benefit all students unless the subtle differences in the nature of classrooms in different EMI schools are taken into consideration. The teaching and learning processes are perhaps different in different EMI science classrooms based on the extent to which EMI is used during lessons. Whether full EMI or a mixed mode of

English and Chinese is adopted in the classrooms is essential to the findings of this research. Instead of determining the learning outcomes of the students, this study investigated the features of classroom interaction in these different EMI settings, and it revealed differences that might contribute to the pedagogical implications of the use of an EMI approach.

4 This study adds to the existing empirical research body regarding the quality of EMI teaching in Hong Kong. While the observed students were only from eight secondary schools around Hong Kong, the participants were carefully selected to be appropriate representatives of Hong Kong students in general regarding their former English-learning experiences, academic achievements during their pre-secondary education years, as well as their socioeconomic background.

10.7.2 *Methodological contributions*

This study was derived from previous research of EMI classroom interactions in the context of Hong Kong (Lo and Macaro, 2012). When analysing classroom interactions, the way that science was taught and its inter-relationship with language-learning issues were considered. This study focused on the way that teachers and students interact with each other using L1 and L2 in the EMI classroom, extending the long history of IRF analysis to explore current language and teaching issues in Hong Kong. In Chapter 5, the two in-depth IRF analyses of student–teacher interactions showed the development of students' thematic understandings in science, providing insights into the connections between macro-level (e.g., different communicative approaches used in science classrooms) and micro-level (e.g., IRF patterns) interactional patterns. Whether teachers selected a student's answer and further expanded on it for discussion was perhaps a determining factor in shaping what Mortimer and Scott (2003) identified as different communicative approaches in science classrooms.

Moreover, this study takes a deeper look into the current existing teaching methods in content learning in secondary schools, including the important tool of the teacher's modified input in developing students' thematic understandings in science. Instead of emphasising the memorisation of scientific vocabulary in L2 to remember certain science ideas, the teacher's modified input facilitates students in relating newly learned concepts to existing thematic patterns, which mediate the concepts they already know or are familiar with. Students are therefore encouraged to understand the concepts in connection with previous knowledge. Science teachers with higher language awareness were found to be more capable of making new science concepts more understandable and accessible to students in L1, mediating abstract concepts to students using their everyday language.

10.7.3 *Theoretical contributions*

As per the interaction hypothesis suggested by Long (1983), learners are more likely to be able to produce more L2 through interactions with teachers when the former is exposed to more comprehensible input. In turn, teachers can give them more modified input as well as corrective feedback, which ultimately contributes to the L2 learning of the students. The gaps among students' language proficiency aside, this study found that skillful teachers are able to mediate science concepts through modified input. This offers interactional scaffolding (e.g., extended IRF patterns) for students to communicate science ideas with the appropriate scientific terms. The data collected also showed that late–partial EMI teachers were more likely to adapt the EMI policy to meet their teaching needs, opting for partial EMI teaching while using Cantonese as the matrix language within which they embed English terms. It was also observed that teachers with a higher level of language awareness used more strategies to promote L2 interaction and production, including code-switching of Cantonese and English, using mind-mapping, encouraging group discussions, asking higher-order questions, and using a range of science questions that would lead to conceptual development in science (Yip, 2004). It may be advisable for science teachers to enhance their language awareness through training and preparation in order to make science content more accessible for students in L2 to pay more attention to their students' language difficulties.

The social context that teachers create through their modified input and interactions with students allows learners to focus and reflect on the language meanings and forms that they have learned and are learning in class. The data collected in this study shows that teacher–student interactions can improve the L2 and science content development of students, particularly when teachers provide modified input in extended IRF exchanged with students.

Mortimer and Scott's (2003) insight that science learning consists of co-constructing meaning through language is reflected by the examples in Chapter 5 of this study. The observations from the four biology classrooms suggested that the co-construction of meaning may take place through the dialogue collaboratively (as in this example, through the extended IRF exchange) between teachers and students, or monologically by a teacher through their modified talk. It is difficult to examine the ability of students to construct meaning by themselves using the expected lexico-grammatical resources in L2 scientific discourse if their science teachers do not make science content more accessible in L2 or pay attention to the students' L2 needs.

According to Johnson and Swain's (1994) argument, students in late immersion programmes face more language difficulties since their existing L2 proficiency level is yet to meet the required threshold level, while their teachers' modified input will respond to the students' L2 challenges. Such modified input can reduce the gap in L2 content learning when students are given sufficient L2 learning opportunities with 'pushed output' on their language production. Teachers' modified input can help enhance the students' L2 proficiency for content learning in subjects such as science, thereby closing the gap between the way that students conceptualise scientific knowledge and what they are able to understand and articulate in English.

It has also been proven that the use of L1 in EMI classrooms can assist L2 learning pedagogically (Lin, 2006; Probyn, 2001, 2009). The data collected in this study supports Lo's (2015) claim that most Hong Kong content-subject teachers took their students' English abilities into account and made thoughtful use of Cantonese to facilitate their learning. In this study, students in both early–full and late–partial EMI schools valued the importance of L1 highly in learning science through English. In the teacher and student questionnaires, the groups from late–partial EMI schools indicated that more Cantonese was used in science teaching activities, especially by teachers when students could not follow in English. This finding perhaps indicates the role of L1 in facilitating teaching and learning in EMI classrooms, and it provides suggestions for optimal L1 usage in supporting students' language development in science. For example, when teachers are explaining a certain scientific concept and L1 is used to facilitate their explanation to make the ideas more understandable for students who have lower English abilities. The study also provides observational evidence of the choices of teachers and students regarding their instructional language, including the use of code-switching across a range of science activities in classrooms. In the interviews conducted, both teachers and students expressed their views on the way in which they moved between Cantonese and English in order to develop content and language, coping with the language demands in EMI classrooms. The data collected on the extent to which code-switching was used by both teachers and students in different EMI classrooms may shed light on the way that the practical use of L1 can help develop students' language abilities so that they can meet the L2 learning requirement as well as master the science content.

10.7.4   *Pedagogical contributions*

This study adds to existing knowledge about the effect of EMI in Hong Kong schools in three crucial ways. First, further pedagogical evidence is provided from the data collected that content learning constituted by different types of EMI programmes (early–full and late–partial EMI schools) can influence various areas of English learning, which include classroom interactions and the attitudes towards EMI learning and teaching. The research also shows that one important factor in evaluating the quality of L2 teaching is the close-up study of the teaching and learning processes in EMI classrooms.

Second, this study suggests some characteristics of EMI teaching that may benefit the language development of learners. Such characteristics include an increase in the frequency of IRF exchanges, teachers' modified input in scaffolding, more opportunities for students to respond to teachers' questions, teachers' feedback, and an increase in opportunities for students to acquire the language required for scientific learning. When helping EMI students acquire both content and language learning in science subjects, the above characteristics are highly effective, helping students develop more positive beliefs about the values of learning as well as stronger self-concepts about their ability to succeed in science learning.

10.8   Recommendations for future research

This study has expanded on some complicated issues of the current MOI situation in Hong Kong secondary schools, including classroom interactions in both early–full and late–partial EMI science classrooms. There is empirical evidence of the effect on the teaching and learning processes of the implementation of the fine-tuning policy which was first introduced forthteen years ago. Previous research has overlooked the implications of the transition to a different MOI in secondary schools.

A longitudinal and comparative approach would be efficient when determining whether schools should opt for a binary decision on their MOI, given the fact that numerous language challenges are found in both early–full and late–partial EMI schools. Future studies may usefully explore whether all secondary schools in Hong Kong should adopt the same EMI policy and whether adopting such an EMI policy would be appropriate for students with rather limited English abilities and experiences in EMI. Future studies should also investigate whether there should be various EMI programmes (e.g., partial EMI groups and transitional EMI groups with more L1 use) in order to help students gradually prepare for full EMI immersion at university in Hong Kong. Future research should also include measures of the impact of EMI and the role of L1 in both language and content-learning outcomes when evaluating different EMI programmes. A longitudinal study should be conducted to investigate whether the adoption of EMI can increase the academic performance of students before and after EMI instruction, providing empirical evidence on the impact of EMI on the actual English development of students in content-learning subjects.

It is also suggested that future research may explore the possibility of alternative EMI transitional programmes designed in order to achieve optimal use of L1 when preparing students for full EMI immersion. It is important to examine the function and significance of L1, as well as how EMI in L1 helps students overcome their language barriers when learning English. The findings from this study may contribute to future discussions of MOI policymaking, one possible implication being the consideration of relaxing the government's 'one language for teaching' policy (either entirely English or entirely Chinese) to allow mixed-mode instruction during the transitional period for the late–partial EMI schools. It is also suggested for future research to explore the conditions and factors involved when establishing these different EMI programmes.

10.9   Concluding remarks and suggestions

This study began by highlighting the fine-tuning language policy, adopted by the Hong Kong government in 2010. The aim of the policy was to encourage schools to adopt English as a MOI for content learning in secondary schools, thereby creating more opportunities for students with low English proficiency to be in an English-learning environment and gradually increasing their exposure and ability in the language. Under the revised policy, most CMI schools have opted for EMI programmes and have become what was referred to in this study as 'late–partial EMI schools'. Although the government has stressed the aim of maximising the educational value brought about by the change of MOI, the reality after the implementation of the policy appears to be different. Many students are faced with language challenges due to the increased amount of EMI teaching, leading to pedagogical and practical issues in content-subject classrooms. The data collected from questionnaires and interviews in this study demonstrated the above-mentioned issues.

The data collected regarding teachers' and students' views on the teaching and learning processes in EMI contexts highlighted the pedagogical and practical challenges that occur while implementing EMI in science classrooms. Many of those challenges are related to students' language skills and teachers' language awareness when addressing their students' language problems. Possible methods and coping strategies are suggested by both teachers and students to address the aforementioned language challenges, particularly for students who are in late–partial EMI schools. Since these students have limited exposure to EMI and may have lower English abilities, they often struggle when learning most content subjects using English. The data collected from interviews and questionnaires in this study revealed the in-class instructional language was not solely English. Teachers tend to switch between L1 and L2 when conducting science activities in class, which reflects that, in practice, most EMI schools have to form their own school-based regulations in order to facilitate their students' learning in a way that best suits the situation (Chan, 2014). This suggests that there is a discrepancy between policymakers' initial aims and the everyday practice of the fine-tuning policy.

The observational data collected in this study revealed that the methods used by teachers to interact with their students in early–full and late–partial EMI science classrooms were different to a small extent, mainly depending on the extent of the implementation of EMI—in other words, whether the class was fully EMI or whether a mixed mode (using both English and Cantonese) was in place. This study found that teachers from different school groups used different methods when teaching in EMI classrooms, with examples provided by the data that inform practical pedagogical suggestions for optimal use of L1 to support students' language development in science subjects. The choice of instructional language is proven in this study to play an important role regarding the number of student initiations and responses, as well as the teachers' questioning techniques, indicated by the number of higher-order questions used as well as the amount of direct feedback provided to students. Regarding the possibility of having evidence that the use of EMI enhances students' content knowledge, the late–partial EMI schools provided empirical proof that environmental conditions (e.g., more L1 use, the use of higher-order questions, more student initiations and responses in the form of extended IRF sequences, less direct feedback to students) are crucial for deeper learning in content subjects, especially for teachers to introduce complex science ideas to their students in order for the latter to comprehend more easily.

Although this study does not directly provide an answer as to whether there is adequate evidence that the use of EMI would enhance students' English proficiency, it has revealed through its analysis and observations of classroom interactions that there is a lack of language support in content learning, which perhaps represents a discrepancy between the aims of the MOI policy and its actual effect on Hong Kong students. The lesson transcripts in this study revealed no incidents in which EMI science teachers and students negotiated the meaning of lexical terms, leading to further L2 learning. Regarding students' L2 language production, students in earl–full EMI schools did not appear to be able to interact with their teachers as freely and expressively as their counterparts in the late–partial EMI schools. Since the former group had experienced EMI for over three years, they might have become accustomed to the EMI teaching methods used in classes. The lack of teacher support and the lack of 'pushed' or 'modified' output from students found in both groups of teachers is perhaps the main challenge in achieving the goal of improving EMI students' English skills, both receptively and productively. This echoes Wannagat's (2007) view that the role of language in the learning process was not clearly amplified in the EMI curriculum, despite students being more exposed to an English-learning environment in EMI classrooms compared with those in Content and Language Integrated Learning (CLIL) classrooms.

The present MOI policy seems to allow a greater extent of flexibility for schools to decide on their own MOI regulations by aiming to reduce the segregation between CMI and EMI streams. However, such flexibility does not necessarily imply solutions to all problems. The government was expected to reveal its assessment of the progress of the fine-tuning MOI policy in 2017, seven years after its implementation, and it should provide successful case studies of effective implementations of the EMI policy at the secondary classroom level, which should include the recognition of frontline teachers' decisions regarding practical teaching interventions, as well as the approval of teachers' judicious use of L1.

Instead of promoting a clear-cut binary policy for EMI and non-EMI in schools, the government is advised to provide clearer guidelines and instructions on the types of teaching methods that could be implemented to handle problems related to the realities of EMI-adopting schools, including the language challenges that students are facing. It is crucial to recognise the different forms of EMI programmes in most EMI secondary schools and to consider a transitional period for both teachers and students in late–partial EMI schools that would allow them to get accustomed to the new teaching and learning environment. For example, the teachers and students in early–full and late–particle EMI schools in this study had different attitudes towards the effects of EMI. Such differences in perceptions at different stages of EMI implementation should be noticed while constructing adequate strategies to account for their individual language challenges.

The EMI policy should be complemented with professional development programmes for frontline EMI teachers so that teachers could be equipped with adequate and necessary pedagogical skills to integrate both content and language teaching (Othman & Saat, 2009). In these programmes, it is suggested for EMI teachers to be trained for English communication skills enhancement, as well as with the specific scaffolding skills needed to deliver content and language to students of different levels through English. Science teachers should be trained on the use of questioning techniques which were suggested by Yip (1999). In order to integrate both content and language when planning EMI lessons, teachers should also be trained in skills and methods such as the ability to unpack scientific ideas and co-construct meaning with students (Rose & Martin, 2010), the code-switching skills to aid scaffolding (Macaro, 2001), and the pedagogical strategies to promote more optimal and productive use of English.

The aforementioned teacher training programmes should be effective in facilitating science teachers in developing their identity as CLIL teachers (i.e., teachers of both academic content and academic language) (Coyle et al., 2010). They should be able to recognise the differences between everyday language and disciplinary language and be able to switch between two languages, instead of teaching techni-cal-scientific phrases solely in L2 (Yip et al., 2003). Schools should also provide a more supportive system for in-service EMI teachers in the early stages of EMI implementation, offering school-based subject matter development. It is believed that with a collaborative team of all science teachers as well as the internal and external support of staff development, innovations such as new teaching interventions for a conceptual change in the EMI classroom could be facilitated (Lo, 2014; Yip, 1999).

10.10 Summary

In conclusion, this study has yielded the following arguments:

1 The extent of the similarities among classroom interactions in early–full and late–partial EMI schools is limited to the macro-level of interaction patterns (with factors such as percentages of interaction time, distribution of time between teacher and student talk, and frequency of pedagogical functions);
2 The different degrees of EMI adoption between early–full and late–partial EMI schools shape the nature of classroom interactions solely at the micro-level. For instance, in the late–partial EMI schools, there are generally more (but shorter) student initiation and responses but also more higher-order questions used by teachers and less direct feedback to students;
3 There was a small decrease in the percentage of interaction time and the maximum length of student turns, as well as a small increase in L1 use, from Grades 10 to 11.
4 Teachers' modified output in classroom interactions plays an important role in developing students' content knowledge and language competence;
5 Empirical data about the role of L1 in aiding learning and teaching in different types of EMI classrooms provide examples of optimal L1 use in supporting students' language development in science;
6 Although teachers in both groups of EMI schools shared similar views regarding their overall EMI teaching experiences, they differed in their opinions on which English language skills are the most important and in their levels of language awareness. Early–full EMI teachers are more aware in terms of the English language and believe that English skills are crucial in the EMI classroom;
7 Students in both groups of EMI schools generally welcome the adoption of EMI at school. However, those from late–partial EMI schools see EMI as an opportunity to improve their English proficiency, while those in early–full EMI schools believe that EMI demotivates their science learning.

This study answered the research questions using the data collected by means of questionnaires, semi-structured interviews, and classroom observations. The approach taken in this study echoes the claim that it is crucial to gather data from multiple sources in order to have a thorough understanding of the nature of patterns in classroom interactions. This study has suggested that a detailed analysis of interaction can be of great help for in-service EMI teachers when developing their understanding of the complexities of science and language teaching. This study has therefore a valuable role regarding the formation of future EMI pedagogy and EMI teacher training. This book has elucidated the different methods to enhance the quality of instructional practices and effective scaffolding in EMI classrooms in Hong Kong and similar contexts around the globe, with collected data and proof-based, pedagogically focused analysis on teacher–classroom interactions in Hong Kong.

References

Chan, J. Y. H. (2014). Fine-tuning language policy in Hong Kong education: Stakeholders' perceptions, practices and challenges. *Language and Education, 28*(5), 459–476. http://doi.org/10.1080/09500782.2014.904872

Coyle, D., Hood, P., & Marsh, D. (2010). *CLIL: Content and Language Integrated Learning*. Cambridge University Press.

Johnson, R. K., & Swain, M. (1994). From core to content: Bridging the L2 proficiency gap in late immersion. *Language and Education, 8*(4), 211–229.

Lin, A. M., (2006). Beyond linguistic purism in language-in-education policy and practice: Exploring bilingual pedagogies in a Hong Kong science classroom. *Language and Education, 20*(4), 287–305. http://doi.org/10.2167/le643.0

Lo, Y. Y. (2015). How much L1 is too much? Teachers' language use in response to students' abilities and classroom interaction in Content and Language Integrated Learning. *International Journal of Bilingual Education and Bilingualism, 18*(3), 270–288.

Lo, Y. Y., & Macaro, E. (2012). The medium of instruction and classroom interaction: Evidence from Hong Kong secondary schools. *International Journal of Bilingual Education and Bilingualism, 15*(1), 29–52.

Long, M. H. (1983). Native speaker/non-native speaker conversation and the negotiation of comprehensible input. *Applied Linguistics, 4*(2), 126–141.

Macaro, E. (2001). Analysing student teachers' codeswitching in foreign language classrooms: Theories and decision making. *The Modern Language Journal, 85*(4), 531–548. http://doi.org/10.1111/0026-7902.00124

Mortimer, E., & Scott, P. (2003). *Meaning making in secondary science classrooms*. Berkshire: England: Open University Press.

Othman, J., & Mohd. Saat, R. (2009). Challenges of using English as a Medium of Instruction: Pre-service science teachers' perspective. *The Asia-Pacific Education Researcher, 18*(2), 307–316.

Poon, A. Y. K. (2013). Will the new fine-tuning medium-of-instruction policy alleviate the threats of dominance of English-medium instruction in Hong Kong?. *Current Issues in Language Planning, 14*(1), 34–51.

Probyn, M. (2001). Teachers voices: Teachers reflections on learning and teaching through the medium of English as an additional language in South Africa. *International Journal of Bilingual Education and Bilingualism, 4*(4), 249–266. http://doi.org/10.1080/13670050108667731

Probyn, M. (2006). Language and learning science in South Africa. *Language and Education, 20*(5), 391–414. http://doi.org/10.2167/le554.0

Probyn, M. (2009). 'Smuggling the vernacular into the classroom': Conflicts and tensions in classroom codeswitching in township/rural schools in South Africa. *International Journal of Bilingual Education and Bilingualism, 12*(2), 123–136. http://doi.org/10.1080/13670050802153137

Rose, D., & Martin, J. R. (2012). *Learning to write, reading to learn: Genre, knowledge and pedagogy in the Sydney School.* Equinox.

Wannagat, U. (2007). Learning through L2–Content and Language Integrated Learning (CLIL) and English as Medium of Instruction (EMI). *International Journal of Bilingual Education and Bilingualism, 10*(5), 663–682.

Yip, D. Y. (1999). Implications of students' questions for science teaching. *School Science Review, 81*(294), 49–53.

Yip, D. Y. (2004). Questioning skills for conceptual change in science instruction. *Journal of Biological Education, 38*(2), 76–83.

Yip, D. Y., Tsang, W. K., & Cheung, S. P. (2003). Evaluation of the effects of medium of instruction on the science learning of Hong Kong secondary students: Performance on the science achievement test. *Bilingual Research Journal, 27*(2), 295–331. http://doi.org/10.1080/15235882.2003.10162808

Index

Note: **Bold** page numbers refer to tables; *italic* page numbers refer to figures and page numbers followed by "n" denote endnotes.